Introduction to Raman Spectroscopy and Its Applications

Carlos Vargas Hernández

Introduction to Raman Spectroscopy and Its Applications

Springer

Carlos Vargas Hernández
Universidad Nacional de Colombia
Sede Manizales, Colombia

Facultad de Ciencias Exactas y Naturales
Departamento de Física y Química
Manizales, Colombia

ISBN 978-3-031-77550-5 ISBN 978-3-031-77551-2 (eBook)
https://doi.org/10.1007/978-3-031-77551-2

This Springer imprint is published by the registered company Springer Nature Switzerland AG
The registered company address is: Gewerbestrasse 11, 6330 Cham, Switzerland

I dedicate this work to:
The Memory of my Father,
Carlos Aristobulo Vargas Serrano,

And my mother *Gladiz Hernández,*

To my family: *My wife, Sandra Patricia, who has supported me throughout my career, and my children, Maria Fernanda, Carlos Andres, and Juan Felipe.*

Preface

Within physical processes, the interaction of electromagnetic radiation with materials is a significant phenomenon of great relevance and importance. This type of interaction allows for the evaluation of the structure and internal dynamics of the components of matter. A wide range of parameters is used to define and classify materials from an optical perspective, grouping their behavior into three major processes: reflection, propagation in the medium, and subsequent transmission.

Reflection involves the discontinuity and participation of two media, each characterized by a refractive index. In electromagnetic wave propagation (EW), two major processes are considered: elastic scattering and inelastic scattering. During the propagation of the EW, absorption occurs, and the frequency of the EW may resonate with the atomic levels within the material, producing an attenuation of the transmitted intensity of the EW that influences the surface.

The importance of studying the optical properties of materials lies in their potential use as optoelectronic devices. These devices have revolutionized the automation and automatic control industry, the communications industry, and the advancement of other technologies. Due to this significant development, it is vital to understand the optical behavior of materials. In this text, among the wide range of optical properties, we will focus on their vibrational and rotational behavior, analyzed through Raman spectroscopy.

Raman spectroscopy has been used since its discovery in the 1920s and today constitutes a powerful technique for quantitatively analyzing materials. The Raman technique is based on the inelastic scattering of light. In earlier times, mercury arc lamps, high-power sources, were used as excitation sources. Long signal accumulation times were needed to achieve a good signal-to-noise ratio in the initial stages of its development. The strengthening of the Raman technique increased significantly in the 1960s with the development of the laser, which constitutes a high-intensity monochromatic source. Around 1980, the Raman technique began to acquire its high potential with advances in photodetector technology, filters, diffraction gratings, and laser diodes.

The central characteristic of the Raman technique is its use for the analysis of the molecular structure of materials, and it is considered a local and, in principle, non-destructive analysis technique. It is a spectroscopic technique used for the analysis of materials in gaseous, liquid, and solid states.

Currently, a modification of the Raman technique is enhanced in nanotechnology applications through the technique known as Surface-Enhanced Raman Scattering (SERS), where metallic nanoparticles are used as an intermediary mechanism to amplify the laser signal through coupling with surface plasmons.

Manizales, Colombia
July 2024

Carlos Vargas Hernández

Acknowledgements I would like to express my gratitude to the National University of Colombia, Manizales Campus, and to the students who have taken the Raman spectroscopy course and suggested the inclusion of essential topics for their professional, research, and academic development.

Introduction

The text is organized into three main parts. Part I covers the theoretical foundations of Raman spectroscopy, including parameters, perturbation theories, normal vibration modes, and the application of group theory to specific structures. It also discusses modifying the Raman technique in the Surface-Enhanced Raman Scattering (SERS) modality. Part II describes the Raman technique and SERS spectroscopy in detail. Part III presents characteristic Raman spectra of various materials measured by the author, along with a basic analysis of their molecular structures.

This book is based on the author's extensive notes, research, and over a decade of teaching Raman spectroscopy courses. Both academic and industrial spectroscopies are crucial due to their analytical applications, offering powerful tools for material study and diagnosis. These techniques provide both quantitative and qualitative insights. The non-invasive nature of optical spectroscopies, combined with advancements in monochromatic sources and high-performance, high-resolution detectors, has led to more versatile and portable instruments. Recent developments in data acquisition, statistical analysis, and signal activation methods—such as SERS—have improved acquisition times and signal-to-noise ratios, enabling high-quality spectra even from deficient sample concentrations.

The text is aimed at final-year science and engineering students and is designed with clear mathematical explanations and detailed steps to help readers grasp the mathematical concepts. It serves as an introduction to Raman spectroscopy, laying the foundation for understanding more complex models in specialized literature.

This book provides a first step into the rapidly evolving field of optical techniques spurred by advancements in optoelectronic and control devices. It covers a broad topic, but to avoid overwhelming the reader, some aspects are not fully explored here. We hope to address these in future publications.

Aspects of the Book

The book is structured into three main parts:

Comprehensive Theoretical Foundations
This part introduces the theoretical fundamentals of Raman spectroscopy, including optical parameters, perturbation theories, normal vibration modes, and the application of group theory to specific structures. It also covers modifications to the Raman technique, specifically the Surface-Enhanced Raman Scattering (SERS) modality.

Depth and Breadth of Theory
Part I of the book provides an extensive overview of the theoretical foundations of Raman spectroscopy. This comprehensive approach ensures that readers gain a thorough understanding of the underlying principles before delving into practical applications, which sets this book apart from others that may focus more narrowly on either theoretical or practical aspects.

Integration of SERS Modality
Cutting-Edge Techniques: The book uniquely addresses the modification of the Raman technique in the SERS modality, highlighting the latest advancements and applications in this area. By including detailed discussions on SERS, the book provides readers with insights into one of the most innovative and rapidly developing areas in spectroscopy. This focus on advanced techniques offers readers knowledge that is highly relevant to current research and technological trends.

Author's Practical Insights and Data
Original Research and Data: Part III of the book presents characteristic Raman spectra of various materials measured by the author, along with a basic analysis of their molecular structure. This inclusion of original research data and practical insights from over a decade of teaching Raman spectroscopy courses gives readers access to real-world examples and applications. This practical perspective, combined with the author's extensive experience, offers a unique learning resource that is both academically rigorous and practically valuable. These unique aspects ensure that your book provides a balanced, in-depth, and contemporary view of Raman spectroscopy, making it a valuable resource for students, researchers, and practitioners in the field.

The Book Addresses the Following Fields of Science or Disciplines

Optical Physics

The book delves into the interaction of light with materials, focusing on Raman spectroscopy and its theoretical foundations, as well as, the SERS technique and their applications. This includes the study of light scattering, vibrational modes, and the optical properties of materials.

Materials Science

It covers the analysis of the molecular structure and properties of various materials through Raman spectroscopy. This field involves understanding the composition, structure, and performance of materials, particularly in relation to their optoelectronic properties.

Analytical Chemistry

The book explores the use of Raman spectroscopy as a powerful analytical tool for qualitative and quantitative analysis. It discusses techniques for identifying chemical compounds and their concentrations in different states of matter (solid, liquid, and gas).

Nanotechnology

It addresses advanced applications of Raman spectroscopy in nanotechnology, specifically through Surface-Enhanced Raman Scattering (SERS). This involves the use of metallic nanoparticles to amplify Raman signals, crucial for analyzing materials at the nanoscale.

These fields highlight the interdisciplinary nature of the book, bridging concepts and applications across various scientific and engineering domains.

Contents

Part I Theoretical Foundations

1 Optical Parameters 3
1.1 Electromagnetic Spectrum 8
1.2 Absorption Coefficient and Penetration Length 14
1.3 Units for Energy 18
1.3.1 Wavenumber 19
1.4 Estimation of Rotational and Vibrational Parameters 20
1.4.1 Rotational Energy of Molecules 25
1.4.2 Vibrational Energy of the Molecules 30
1.5 Proposed Problems 32
References 33

2 Quantum Mechanics Theory 35
2.1 Postulates of Quantum Mechanics 36
2.1.1 Postulate 1 36
2.1.2 Postulate 2 36
2.2 Hermitian Operators 37
2.3 Theorems Associated with Hermitian Operators 39
2.3.1 T1 39
2.3.2 T2 40
2.4 Theorem, Hermitian Operator 40
2.5 Theorem Orthogonality 41
2.6 Degenerate States 41
2.7 Linear Combination of Eigenfunctions 42
2.8 Theorem of Real Value 42
2.9 Algebra of Operators 43
2.9.1 Linear Operators 43
2.9.2 Commutator 44
2.9.3 Eigenfunctions and Eigenvalues 45

2.10 Commutator ... 46
2.10.1 Definition ... 46
2.10.2 Properties of Commutators ... 46
2.11 Proposed Problems ... 51
References ... 51

3 Mathematical Fundamentals of Perturbation Theory ... 53
3.1 Time-Independent Perturbation Theory ... 53
3.1.1 Zero Order ... 55
3.1.2 First Order ... 55
3.1.3 Second Order ... 55
3.1.4 Terms of Order Q ... 55
3.2 Perturbation Theory for Non-Degenerate Energy Levels ... 56
3.2.1 First-Order Energy Correction ... 56
3.2.2 First-Order Correction of Eigenstates ... 57
3.2.3 First-Order Wavefunction Correction in the First State ... 58
3.2.4 Second-Order Energy Corrections ... 59
3.2.5 *Example* The Stark Effect ... 59
3.2.6 *Example* Harmonic Oscillator ... 61
3.3 Time-Dependent Perturbation Theory ... 63
3.3.1 Example ... 67
3.4 Variational Method ... 69
3.4.1 Example of the Variational Method ... 72
3.5 Proposed Problems ... 78
References ... 79

4 Harmonic Crystals ... 81
4.1 Lagrangian Equation ... 81
4.2 Normal Coordinates ... 82
4.3 Planar Equilateral Triangle ... 87
4.4 Matrix Method ... 101
4.4.1 Applications ... 101
4.4.2 Generalized Matrix Equation ... 110
4.5 Other Coordinates ... 116
4.5.1 Concept of Phonons ... 117
4.6 Proposed Problems ... 117
References ... 118

5 Raman Spectroscopy ... 119
5.1 Theoretical Foundations of Raman Spectroscopy ... 119
5.2 Historical Review of the Raman Spectroscopy ... 121
5.3 Classical Theory of the Raman Effect ... 121
5.4 Selection Rules ... 125

5.5 Relation Between Stokes and Anti-Stokes Intensities 127
5.6 Raman Cross Section 129
5.7 Quantum Theory of the Raman Effect 130
5.8 Relation Between Symmetry, Selection Rules, and the Intensities of Stokes and Anti-Stokes Lines 132
5.8.1 Example of a Raman Spectroscopy Calculation 132
Reference 133

Part II SERS, Group Theory, and Symmetries

6 SERS Theory 137
6.1 Theoretical Foundations 137
6.2 SERS Theory 137
6.3 Plasmons and Electromagnetic Theory 139
6.3.1 Electromagnetic Theory in Metals 141
6.3.2 Free Electron Gas Model 144
6.3.3 Localized Surface Plasmons 150
References 159

7 Group Theory 161
7.1 Group Theory 162
7.2 Representation of Geometric Transformations by Matrices 163
7.2.1 The Identity E 164
7.3 Rearrangement Theorem 167
7.4 Symmetry 168
7.5 Notation 169
7.6 Cosets 170
7.7 Conjugate Elements and Class Structure 171
7.8 Conjugate Elements 172
7.8.1 Representation Theory 172
7.8.2 Matrix Representation 172
7.8.3 Character of a Representation 173
7.8.4 Dimensionality Theorem 173
7.8.5 Construction of the Character Table 173
7.9 Reducible Representation 175
7.9.1 Regular Representation 175
7.9.2 Cayley Square Table D_4 176
7.9.3 Projection Operators 177
7.10 Application of Projection Operators 181
7.10.1 Triatomic Molecule with Equilateral Geometry 181
7.11 Proposed Problems 183
References 184

8 Symmetries and Quantum Mechanics ... 185
8.1 Symmetry and Quantum Mechanics Applied to Molecules ... 186
8.1.1 Symmetry in Molecular Quantum Mechanics ... 186
8.1.2 Quantum Mechanical Operators and Symmetry ... 187
8.1.3 Selection Rules in Spectroscopy ... 187
8.1.4 The Hamiltonian and Symmetry ... 188
8.1.5 Applications in Molecular Quantum Mechanics ... 188

Part III SERS Spectroscopy Technique

9 Raman Technique ... 193
9.1 Introduction ... 193
9.2 Definition of Terms in Raman Terminology ... 193
9.3 Raman Technique ... 194
9.3.1 FT-Raman Spectroscopy ... 194
9.3.2 Differences and Similarities Between Raman and FTIR Spectroscopy ... 195
9.3.3 Principle of the Raman Technique ... 197
9.3.4 Instrumentation Used for Dispersive Raman Technique ... 198
9.3.5 Light Sources Used in Sample Excitation ... 199
9.3.6 Detectors ... 199
9.3.7 Monochromator ... 200
9.3.8 Some Types of Monochromators Used in Spectroscopy ... 201
References ... 202

10 SERS Spectroscopy Technique ... 203
10.1 Principle of SERS ... 203
10.2 SERS Substrates ... 204
10.3 Applications of SERS ... 204
10.3.1 Advantages of SERS ... 205
10.3.2 Challenges and Future Directions ... 205
10.4 Equations for SERS ... 205
10.4.1 Electromagnetic Enhancement Factor ... 205
10.4.2 SERS Intensity ... 206
References ... 209

11 Raman Spectra of Some Materials ... 211
11.1 Inorganic Materials ... 212
11.1.1 Silicon Calibration Standard ... 212
11.1.2 Zinc Oxide ... 213
11.1.3 Vanadium Oxide ... 216
11.1.4 Zinc Chromite $ZnCr_2O_4$ ... 217
11.1.5 KDP ... 218
11.1.6 Gallium Arsenide ... 219

11.1.7 GaAs/GaAs. GaAs Film Growth on a GaAs Substrate (i.e., GaAs/GaAs) 221
11.1.8 Calcite 222
11.2 Organic Materials 223
11.2.1 Cellulose 223
11.2.2 Epoxy Resin 224
11.2.3 TMAB Diamine 224
11.3 Other Materials 225
11.3.1 Hair 225
11.3.2 Bone 228
11.3.3 Snail Shell 230
11.3.4 Water 231
11.3.5 PEO 232
11.3.6 Styrofoam 233
11.3.7 DNA 235
References 239

Bibliography 241

Index 243

About the Author

Carlos Vargas Hernández is a Full Professor at the National University of Colombia, specializing in physics, mathematics, and electronics at the undergraduate, masters, and doctoral levels. His academic and research career is distinguished by a strong focus on the optical and optoelectronic properties of materials. He is the Director of the Laboratory of Physical Properties of Materials at the National University of Colombia and was previously the Director of the Physics Department. His research has concentrated on the characterization of materials mixed with metallic nanoparticles for biotechnological applications using high-resolution Raman spectroscopy systems. His academic background includes degrees in Physics and Electronic Engineering, along with three master's degrees: Master of Science in Physics, Master of Industrial Automation, and Master of Electrical Engineering. He has earned two doctorates: Doctor of Science in Physics from CINVESTAV-IPN, Mexico, and Doctor of Engineering from the National University of Colombia. He completed a postdoctoral fellowship at the University of Texas and was a Visiting Professor at the University of Toronto. His current research focuses on the study, design, and growth of advanced materials such as transistors and biotransistors activated through redox reactions. His work involves the theoretical study of ultrathin quantum wells and quantum dots as transistor gate drivers. His material growth techniques include molecular beam epitaxy, chemical bath deposition, microwave-assisted solid reaction, and microwave-assisted chemical bath.

Acronyms

CARS	Coherent Anti-Stokes Raman Spectroscopy
EF	Enhancement Factor
EW	Electromagnetic Wave
FT-RS	Fourier Transform Raman Spectroscopy
HRS	Hyper Raman Spectroscopy
HWRS	High Wavenumber Raman Spectroscopy
LPSR	Localized Surface Plasmon Resonance (relevant to SERS)
NIR	Near-Infrared Region
NRRS-VNIR	Non-resonant Raman Spectroscopy in Visible and Near-Infrared
NRS	Non-linear Raman Spectroscopy
RIM	Raman Imaging Microscopy
RRS	Resonant Raman Spectroscopy
RRS-UV	Resonant Raman Spectroscopy in Ultraviolet
RS	Raman Spectroscopy
SECARS	Surface-Enhanced Coherent Anti-Stokes Raman Spectroscopy
SERS	Surface-Enhanced Raman Spectroscopy
SRS	Stimulated Raman Scattering
TERS	Tip-Enhanced Raman Spectroscopy
TRRS	Time-Resolved Raman Spectroscopy

Part I
Theoretical Foundations

Part I of the book is structured to provide information on the most commonly used optical parameters in Raman spectroscopy and chapters that offer the necessary foundation to understand this technique and its associated methods. Therefore, we have included chapters on the optical parameters used in spectroscopy, along with basic information on the theory of quantum mechanics, its foundations, postulates, and representations.

One of the fundamental pillars of modern technology and science is quantum mechanics; therefore, it is crucial to present its postulates, theorems, and representations through the respective operators. Additionally, we have included the mathematical foundations of perturbation theory, a powerful methodology that allows us to obtain results from weakly perturbed systems and correct their respective energies.

The chapter on harmonic crystals aims to apply classical theories, complemented by quantum theories, to obtain approximate results through the model known as quantum mechanics. The theory of harmonic crystals combines symmetries by applying differential equation methods, such as the Lagrange equation and matrix methods. Through matrix representation, the latter allows the use of more systematic and efficient methodologies, facilitating the solution of systems in conjunction with symmetry theories.

Finally, in the last chapter of Part I, the theoretical foundations of Raman spectroscopy are presented. Here, both the classical and quantum models of this technique are explained, and examples are included in each chapter.

Chapter 1
Optical Parameters

Abstract Understanding and accurately measuring optical parameters is essential in spectroscopy because these parameters provide detailed information about the composition, concentration, and properties of materials. This knowledge is vital in various fields, from scientific research to industrial applications, and helps develop and optimize optical technologies. In the topic at hand, such as Raman spectroscopy, knowing parameters such as the refractive index and optical absorption allows us to design and automate experiments, as well as understand both the signals obtained and the respective spectra, which leads to conclusions about the problem under study.

When light strikes a material, phenomena can occur that depend on the spectral region of the wave interacting with the material. These phenomena can be classified as either surface or volume phenomena. At the surface contact, due to the discontinuity between the medium and the material, the electromagnetic wave interacts with the medium through its fields, producing the well-known effects of reflection and transmission. This discontinuity, associated with the different refractive indices present between the sample and the environment, allows for the study of the fraction of incident and transmitted photons, as well as the various phenomena associated with interfacial fields and the characteristic surface reconstructions of each material. With the fraction of light transmitted within the volume of the sample, information can be obtained about the organization of atoms or molecules, defects, symmetry of the structure, etc.

Light is an electromagnetic wave, and as such, it consists of oscillating electric and magnetic fields in phase, allowing it to interact with matter through the charge surrounding the atoms or molecules. In this phenomenological interpretation, the interaction of light with matter can be visualized as a radiation-matter interaction and, from the perspective of modern physics, it is studied as an electron-photon interaction. To understand the effect known as Raman scattering, it is necessary to comprehend how electromagnetic radiation couples and exchanges energy through its electric field with the electrons that form the material it interacts with. Table 1.1 illustrates the region of interest in the electromagnetic spectrum where Raman scattering occurs, approximately between the infrared (vibrations) and microwaves (rotations).

C. Vargas Hernandez, *Introduction to Raman Spectroscopy and Its Applications*,
https://doi.org/10.1007/978-3-031-77551-2_1

Table 1.1 Frequency and range of interaction

Frequency (Hz)	Wavelength (λ)	Radiation name	Involved transition
$10^{20} - 10^{24}$	$10^{-12} - 10^{-16}$ m	Gamma rays	Nuclear
$10^{17} - 10^{20}$	1 nm–1 pm	X-rays	Inner electrons
$10^{15} - 10^{17}$	400–1 nm	Ultraviolet	Outer electrons
$4.310^{14} - 7.510^{14}$	700–400 nm	Visible	Outer electrons
$10^{12} - 10^{14}$	2.5 μ m–700 nm	Infrared	Vibrations
$10^{8} - 10^{12}$	1 mm–2.5 μ m	Microwaves	Rotations
$10^{0} - 10^{8}$	$10^{8} - 1$ m	Radio frequency	Spin reversal

In modern physics, electromagnetic radiation is quantized, and the quantum of radiation is called a photon, where energy, frequency, and wavelength are related to each other [1, 2].

The energy of the photon can be obtained using the equation (1.3):

$$E_{Photon} = hf \tag{1.1}$$

where

- E_{Photon} is the energy of the photon,
- h is Planck's constant ($h = 6.626 \times 10^{-34} J.s$), and
- f is the frequency of the photon.

Alternatively, since the speed of light c is related to the frequency and wavelength (λ) of the photon by the equation $c = \frac{\lambda}{T} = \lambda f$, then

$$E_{Photon} = \frac{hc}{\lambda} \rightarrow E_{Photon} = \frac{1239.84 \text{ eV nm}}{\lambda\ (nm)} \tag{1.2}$$

where,

- c is the speed of light in a vacuum ($c = 3.00 \times 10^{8} \frac{m}{s}$) and
- λ is the wavelength of the photon.

A photon is a fundamental particle representing a quantum of light, and it is the force carrier for the electromagnetic force [4]. Photons are constantly in motion and, in a vacuum, travel at the speed of light. The photons have some characteristics as

Massless: Photons have zero rest mass, which allows them to travel at the speed of light in a vacuum.

Charge: Photons have no electric charge.

Energy and Momentum: Even though photons have no mass, they carry energy and momentum related to their frequency and wavelength. The momentum p is defined by $p = \frac{E_{Photon}}{c}$.

Wave-Particle Duality: Photons exhibit both wave-like and particle-like properties. They can interfere and diffract like waves but also exhibit particle-like behavior in phenomena such as the photoelectric effect, which can be thought of as discrete packets of energy. They manifest depending on how you interact with them.

Polarization: Photons can be polarized, meaning their electromagnetic fields can oscillate in specific directions.

Interaction with Matter: Photons interact with charged particles, such as electrons and protons, through electromagnetic forces [5]. These interactions can result in various phenomena, such as absorption, emission, and scattering.

Understanding photons is crucial for explaining many physical phenomena and technologies, such as lasers, photovoltaic cells, and various forms of spectroscopy, such as Raman spectroscopy. The choice of laser wavelength in Raman spectroscopy depends on the sample being analyzed and the specific requirements of the experiment, such as whether the sample is solid or liquid. Other devices, such as filters and detectors, must be considered for resolution and power.

The type or class of laser used depends on its balance between scattering efficiency and fluorescence interference and its suitability for a wide range of samples, including biological materials and inorganic compounds, whether liquid or solid. The excitation energy is higher for shorter wavelengths and generally provides stronger Raman signals but can induce greater fluorescence. This is because the scattering power is proportional to the inverse of the wavelength raised to the fourth power. Longer wavelengths are preferred to minimize the effects of fluorescence (e.g., 785 nm or 1064 nm), but this depends on the sample type. These wavelengths are ideal for biological samples, dyes, and other fluorescence-prone materials, providing greater penetration into the sample and reduced heating. However, they typically require more sensitive detection equipment due to lower Raman scattering efficiency. It is important to remember that shorter wavelengths produce stronger signals and induce more fluorescence. However, they can also deteriorate the sample due to their high photon energy, which produces greater scattering power and possibly generates more heat. Specific wavelengths may be required for optimal results in biological samples, organic compounds, and certain minerals. Ultimately, the choice of laser wavelength is fundamental to optimizing the results of Raman spectroscopy for different types of samples.

Standard laser wavelengths used in Raman spectroscopy: 473 nm (blue),532 nm (Green), 633 nm (Red), 785 nm (near infrared), 1064 nm (near infrared).

Examples Calculation

Let's calculate the energy of a photon with different wavelengths in nanometers (nm), which is in the visible spectrum of light.

Example For wavelengths between $\lambda = 2.5\mu\text{m}$ and $\lambda = 1\text{mm}$, and using Planck's constant $h = 4.036 \times 10^{-15} eV \cdot s$ and light speed $c = 2.998 \times 10^{8} \frac{m}{s}$ the energies obtained are, respectively:

$$E_{Photon} = \frac{hc}{\lambda} \rightarrow \left| \begin{array}{l} \text{for} \, 1\,\text{mm} \rightarrow E_{Photon} = \frac{hc}{\lambda} = 1.24\ \text{meV} \\ \\ \text{for } 2.5\ \text{um} \rightarrow E_{Photon} = \frac{hc}{\lambda} = 495.99\ \text{meV} \end{array} \right. \quad (1.3)$$

Example What are the energies of three photons with wavelengths $\lambda_1 = 473$ nm, $\lambda_2 = 633$ nm and $\lambda_3 = 1064$ nm, and how are these energies related to their wavelengths?

A useful relation that is used is $h \cdot c = 1.24 \times 10^{3}$ J

$$E_{Photon} = \frac{hc}{\lambda} \rightarrow \left| \begin{array}{l} \text{for} \lambda_1 = 473\,\text{nm} \rightarrow E_{Photon} = \frac{hc}{\lambda_1} = 2.622\ \text{eV} \\ \\ \text{for } \lambda_2 = 633\ \text{nm} \rightarrow E_{Photon} = \frac{hc}{\lambda_2} = 1.959\ \text{eV} \\ \\ \text{for } \lambda_3 = 1064\ \text{nm} \rightarrow E_{Photon} = \frac{hc}{\lambda_3} = 1.165\ \text{eV} \end{array} \right. \quad (1.4)$$

The Table 1.1 indicates that vibrational and rotational transitions occur approximately in an energy range between 1 and 500 meV. For example, to activate the longitudinal optical normal vibration mode of GaAs, around 36 meV is needed, while to activate the bending mode in the water molecule, 420 meV is required.

When light strikes a material, phenomena can occur that are classified as surface or volume phenomena, depending on the region of interaction. The electromagnetic wave, through its fields in contact with the surface, exhibits reflection and transmission effects due to the discontinuity created by the presence of the medium with which it interacts. This discontinuity, associated with different refractive indices, allows for the study of the fraction of photons reflected and transmitted, as well as phenomena related to interfacial fields and surface reconstructions. With the fraction of light transmitted into the volume of the sample, it is possible to obtain information about the organization of atoms, molecules, defects, and symmetry, among other aspects.

The number of photons in a commercial laser pulse or beam depends on several factors, including the wavelength of the laser light, the power output, and the duration of the pulse (for pulsed lasers) or the time over which the power is measured (for continuous wave lasers). To estimate the number of photons, the following equation can be used:

$$\left| \begin{array}{l} P_{Laser} = \dfrac{E_{Laser}}{t} \\ E_{Laser} = N_{Photons} E_{Photons} \\ E_{Photon} = \dfrac{hc}{\lambda_{Laser}} \\ N_{Photons} = \dfrac{E_{Laser}}{A_{Spot} E_{Photons}} \end{array} \right| \rightarrow N_{Photons} = \frac{P_{Laser} t \lambda_{Laser}}{A_{Spot} hc} \quad (1.5)$$

- P_{Laser} is the power of the laser in watts or milliwatts;
- E_{Laser} is the energy of the laser in joules or electronvolts;
- t is the time in seconds (s) (for continuous wave lasers, this would be the duration of the measurement; for pulsed lasers, it would be the pulse duration);
- $E_{Photons}$ is the photon energy in electronvolts;
- λ_{Laser} is the wavelength of the laser device in nanometer;
- $N_{Photons}$ is the number of photons that incident per second; and
- A_{Spot} is the area of the laser spot.

Example Consider a commercial laser with the following specifications:

Wavelength ($\lambda_{Laser=}$ 632.8 nm (Red laser), power $P_{Laser} = 5$ mW (typical for many commercial laser pointers), duration time 1 s (for continuous wave laser over 1 s),and $A_{Spot} = 4\text{mm}^2$. Then, calculate the number of photons as

$$\lambda_{Laser} = 632.8\text{nm} \rightarrow E_{Photon} = 1.959\text{eV} \quad (1.6)$$

as

$$E_{Laser} = P_{Laser} \cdot t \quad (1.7)$$

then

$$n_{Photons} = \frac{E_{Laser}}{A_{Spot} \cdot E_{Photon}} = \frac{P_{Laser} \cdot t}{A_{Spot} \cdot E_{Photon}} \quad (1.8)$$

therefore

$$n_{Photons} = \frac{5\text{mW} \cdot 1\text{s}}{4\text{mm}^2 \times 1.959\text{eV}} = 3.982 \times 10^{15} \cdot \frac{1}{mm^2} \quad (1.9)$$

So, a 5 mW red laser operating for 1 s emits approximately 3.982×10^{15} photons per mm^2. The exact number will vary based on the laser's power and operational time.

1.1 Electromagnetic Spectrum

The interaction between radiation and matter is understood through the use of Maxwell's equations, which in their differential form are expressed as follows [6, 7]:

$$\left| \begin{array}{l} \nabla \cdot \mathbf{E} = \rho_v \\ \nabla \cdot \mathbf{H} = 0 \\ \nabla X\mathbf{E} = -\frac{\mu}{c}\frac{\partial \mathbf{H}}{\partial t} \\ \nabla X\mathbf{H} = \frac{4\pi\sigma}{c}\mathbf{E} + \frac{\varepsilon}{c}\frac{\partial \mathbf{E}}{\partial t} \end{array} \right. \tag{1.10}$$

Additionally, the constitutive relation that links the electric field **E** and the current density **J** in the volume where the electric field is present is expressed as

$$\mathbf{J} = \sigma\mathbf{E} \tag{1.11}$$

where **E**, ρ_v, **H**, μ, c, σ, ε, and **J** are the electric field, the volumetric density of free charge, the magnetic field, the magnetic permeability, the speed of light, the electrical conductivity, the dielectric constant, and the current density, respectively. Let's consider that the material interacting with the electromagnetic radiation is electrically neutral, in which case $\rho_v = 0$.

The following identity of vector calculus states that

$$\vec{\nabla} x \vec{\nabla} x \vec{A} = \vec{\nabla}\left(\vec{\nabla} \cdot \vec{A}\right) - \nabla^2 \vec{A} \tag{1.12}$$

It will use the vector analysis property of the double curl of the electric field to obtain the electromagnetic wave equation.

Using:

$$\vec{\nabla} X \vec{E} = -\frac{\mu}{c}\frac{\partial \mathbf{H}}{\partial t} \tag{1.13}$$

By applying the curl to the previous equation, we obtain

$$\overrightarrow{\nabla} x \overrightarrow{\nabla} x \overrightarrow{E} = -\frac{\mu}{c}\frac{\partial \overrightarrow{\nabla} x\mathbf{H}}{\partial t} = \overrightarrow{\nabla}\left(\overrightarrow{\nabla}\cdot\overrightarrow{E}\right) - \nabla^2\overrightarrow{E} = -\nabla^2\overrightarrow{E} \tag{1.14}$$

This leads us to

$$-\frac{\mu}{c}\frac{\partial \overrightarrow{\nabla} x\mathbf{H}}{\partial t} = \frac{4\pi\sigma}{c}\frac{\partial \overrightarrow{E}}{\partial t} + \frac{\varepsilon}{c}\frac{\partial^2 \overrightarrow{E}}{\partial t^2} \tag{1.15}$$

Given that $\overrightarrow{\nabla}\cdot\overrightarrow{E} = 0$ in the absence of free charges, it simplifies to

$$-\nabla^2\overrightarrow{E} = -\frac{\mu}{c}\frac{\partial \overrightarrow{\nabla} x\mathbf{H}}{\partial t} \tag{1.16}$$

Substituting $\overrightarrow{\nabla} x\mathbf{H} = \frac{4\pi\sigma}{c}\overrightarrow{E} + \frac{\varepsilon}{c}\frac{\partial \overrightarrow{E}}{\partial t}$. It obtains the representation of the electromagnetic wave equation as

$$\nabla^2\overrightarrow{E} = \frac{4\pi\sigma\mu}{c^2}\frac{\partial \overrightarrow{E}}{\partial t} - \frac{\mu\varepsilon}{c^2}\frac{\partial^2 \overrightarrow{E}}{\partial t^2} \tag{1.17}$$

The solution to this second-order differential equation is a plane wave. For practical purposes, let's assume it propagates in the positive x-direction:

$$E_x = E_o e^{-i(\omega t - \eta x)} \tag{1.18}$$

where

$$\eta = \omega/v \tag{1.19}$$

where ω and v are the frequency and the speed of the electromagnetic wave in the propagation medium. Substituting into the wave equation, we get

$$\eta^2 = \frac{\mu\varepsilon\omega^2}{c^2} - i\frac{4\pi\mu\sigma\omega}{c^2} \tag{1.20}$$

$$\frac{1}{v^2} = \frac{\mu\varepsilon}{c^2} - i\frac{4\pi\mu\sigma}{\omega c^2} \tag{1.21}$$

One of the most commonly used parameters in materials is the refractive index N , which is a complex function and can be expressed with a real part n and an imaginary part k, called *coeficiente de extinción*:

$$N = n + ik \tag{1.22}$$

Using the corresponding algebra, we obtain that [8]

$$\begin{aligned} n^2 &= \tfrac{\mu\varepsilon}{2}\left[\left(1+\left(\tfrac{2\sigma}{\varepsilon v}\right)^2\right)^{1/2}+1\right] \\ k^2 &= \tfrac{\mu\varepsilon}{2}\left[\left(1+\left(\tfrac{2\sigma}{\varepsilon v}\right)^2\right)^{1/2}-1\right] \end{aligned} \tag{1.23}$$

Through these functional expressions, the behavior of the material when exposed to electromagnetic radiation is correlated. There are particular cases depending on whether the material is a conductor, semiconductor, or insulator. In the case of insulating materials, we consider $J = 0$ and the previous equations simplify.

For each component of the dielectric function, we obtain

$$\begin{aligned} \varepsilon_1 &= n^2 - k^2 \\ \varepsilon_2 &= 2nk \end{aligned} \tag{1.24}$$

where k is the extinction coefficient and n is the real refractive index. Using (1.22), and equating the real and imaginary parts, in this case, it has

$$\left| \begin{aligned} n^2 - k^2 &= \mu\varepsilon - \tfrac{\mu\sigma}{\omega} \\ nk &= \tfrac{\mu\sigma}{v} \end{aligned} \right. \tag{1.25}$$

From Eq. 1.25, we can determine n and k as follows:

$$n^2 + k^2 = \left[(\mu\varepsilon)^2 + (\frac{2\mu\sigma}{v})^2\right]^{1/2} \tag{1.26}$$

Finally, using (1.25) and (1.26), we obtain

$$\left| \begin{aligned} n &= \sqrt{\tfrac{\mu\varepsilon}{2} + \sqrt{\left(\tfrac{\mu\varepsilon}{2}\right)^2 + \left(\tfrac{2\sigma}{\omega}\right)^2}} \\ k &= \tfrac{\mu\sigma}{2\omega n} \end{aligned} \right. \tag{1.27}$$

If the material is an insulator,$\sigma \to 0$, then $k \to 0$, which means there is no absorption in the material. If we define $\varepsilon = \varepsilon_r\varepsilon_0$ and $\mu = \mu_r\mu_0$, we achieve for each of the components of the dielectric function the following:

$$\left| \begin{aligned} \varepsilon_r &= n^2 \\ \mu_r &= 1 \end{aligned} \right. \tag{1.28}$$

We express the relationship between absorption and extinction coefficients through the following:

$$\alpha = \frac{2\omega k}{c} = \frac{4\pi k}{\lambda} \tag{1.29}$$

Reflectance is found from the Fresnel coefficients, which are established as the ratio between the intensity of the incident and reflected or transmitted electric field. For normal incidence, the reflection coefficient is determined by

$$R = \frac{(n-1)^2 + k^2}{(n+1)^2 + k^2} \tag{1.30}$$

It considers a theoretical treatment of the dispersion phenomenon by free and bound electrons, contributing to the dielectric function. Electromagnetic radiation interacts in many ways with matter; the physical processes by which these interactions occur are called scattering mechanisms. The three most representative scattering mechanisms in a semiconductor, when it interacts with electromagnetic waves, are

Bound electrons: These are electrons bound to the atoms of a crystal, which do not contribute to electrical conduction, and this interaction manifests in the ultraviolet, visible, and part of the near-infrared ranges.

Free electrons or holes: These are electrons that, being free, contribute to electrical conduction, and this interaction manifests in the mid-infrared and part of the far-infrared.

Ions and atoms: They interact with electromagnetic radiation through their ionic charge; this scattering is usually referred to as lattice scattering and manifests in the mid-infrared, far-infrared, and part of the microwave region.

In dielectric materials, the electrons are strongly bound to their constituent atoms but can be displaced by an electric field. According to the above, we propose an oscillator model where the bound electrons oscillate around an equilibrium point under an external force that perturbs them, this being an applied electric field corresponding to the incident electromagnetic radiation. We assume a force proportional to displacement and a friction proportional to velocity. We obtain the following differential equation:

$$m\frac{d^2x}{dt^2} + \gamma\frac{dx}{dt} + \beta x = -eE \tag{1.31}$$

where β and γ are the restoring constant and the damping coefficient, respectively. The natural frequency of electron vibration is defined as $\omega_0 = \sqrt{\frac{\beta}{m}}$. The electric field that causes the perturbation has the form $E = E_0 \exp(i\omega t)$.

The solution of Eq. 1.31 leads to the following expression for displacement:

$$x = \frac{eE_0}{m\left(\left(\omega_0^2 - \omega^2\right) + i\gamma\omega\right)} \exp(i\omega t) \tag{1.32}$$

The electric polarization can be expressed by

$$P = Nex \tag{1.33}$$

Using Eqs. 1.32 and 1.33, it yields

$$P = \frac{Ne^2E_0}{m\left(\left(\omega_0^2 - \omega^2\right) + i\gamma\omega\right)} \exp(i\omega t) \tag{1.34}$$

Rationalizing and using Eq. 1.28, we arrive at the expressions:

$$\varepsilon_r = n^2 = 1 + \frac{Ne^2}{\varepsilon_0 m} \sum_j \frac{1}{\left[\left(\omega_{0j}^2 - \omega^2\right) + i\gamma_j^2\omega\right]} \tag{1.35}$$

Sellmeier obtained the following formula for the variation of the refractive index as a function of the wavelength, for a medium with several absorption bands j, where N is the number of electrons per unit volume and λ_j are the different resonance frequencies:

$$n^2 = 1 + \sum_j^b \left(\frac{e^2}{4\pi\varepsilon_0 m} \left[\frac{N\lambda_j^2\lambda^2}{\left(\lambda^2 - \lambda_j^2\right)} \right] \right) \tag{1.36}$$

The classical model is still incomplete for describing the interaction of electromagnetic radiation with matter. In quantum mechanical theory, the interaction of fields with atoms can be represented as a set of linear oscillators, each with its own resonance frequency corresponding to an optical transition. The magnitude of each oscillator's contribution to the optical parameters is determined by the oscillator strength in each transition. Using Eq. 1.35 and summing over all the oscillators for all the allowed transitions, we obtain

$$\left| \begin{array}{l} n^2 + k - 1 = \sum_j \frac{Ne^2\left(\omega_j^2 - \omega^2\right) f_j}{\varepsilon_0 m\left[\left(\omega_j^2 - \omega^2\right)^2 + \gamma_j^2\omega^2\right]} \\ 2nk = \sum_j \frac{Ne^2\omega\gamma_j f_j}{\varepsilon_0 m\left[\left(\omega_j^2 - \omega^2\right)^2 + \gamma_j^2\omega^2\right]} \end{array} \right. \tag{1.37}$$

In quantum mechanics, polarizability and the dielectric function are obtained in the same way as in the classical Lorentz model. In the following sections, we will consider the interaction of a free atom with an electromagnetic field, assuming the ground state (unperturbed, not excited), so we only need to consider the effects of polarization and absorption, not emission (since it is not perturbed).

The spectral variation of the refractive index can be written as follows using Eq. (16), which is another form of the Sellmeier formula [5]:

$$n^2(\lambda) = 1 + \sum_{i=1}^{N} \frac{B_i \lambda^2}{\left(\lambda^2 - C_i^2\right)} \tag{1.38}$$

where $n(\lambda)$ is the refractive index at wavelength λ, B_i and C_i are empirically determined coefficients specific to the material, and N is the number of terms in the series (usually 1 to 3 terms are sufficient for most materials).

The Sellmeier oscillator refers to a model used in optics to describe the frequency dependence of the refractive index of a material. The Sellmeier equation, which is the core of this model, relates the refractive index to the wavelength of light. This equation is particularly useful for characterizing optical materials over a range of wavelengths and is crucial in designing optical systems such as lenses, prisms, and other components.

Examples

Example fused silica. For a common optical material like fused silica (SiO_2), the Sellmeier coefficients might be provided as

- $B_1 = 0.9661663$
- $B_2 = 0.4079426$
- $B_3 = 0.8974794$
- $C_1 = (0.0684043)^2$
- $C_2 = (0.1162414)^2$
- $C_3 = (9.896161)^2$

What is the refractive index of fused silica at a wavelength of $\lambda = 0.55\mu\text{m}$?

Problem Statement Replace the wavelength and coefficients into the Sellmeier Eq. 1.38, and calculate each term of the series and sum them. In that case, it has

$$n^2(\lambda) = 1 + \frac{B_1 \lambda^2}{\left(\lambda^2 - C_1^2\right)} + \frac{B_2 \lambda^2}{\left(\lambda^2 - C_2^2\right)} + \frac{B_3 \lambda^2}{\left(\lambda^2 - C_3^2\right)} \tag{1.39}$$

then

$$n^2(0.55) = 1 + \frac{0.6961663 \times (0.55)^2}{(0.55)^2 - (0.0684043)^2} + \frac{0.4079426 \times (0.55)^2}{(0.55)^2 - (0.1162414)^2} + \frac{0.8974794 \times (0.55)^2}{(0.55)^2 - (9.896161)^2}$$

$$n(0.55) = \sqrt{2.13} = 1.46 \tag{1.40}$$

1.2 Absorption Coefficient and Penetration Length

One of the crucial parameters in the interaction between radiation and matter is the absorption coefficient α, which can be related to the frequency of the incident radiation ω, the extinction coefficient, and the wavelength λ. It is expressed as

$$\alpha = \frac{2\omega k}{c} = \frac{4\pi k}{\lambda} \tag{1.41}$$

When electromagnetic radiation interacts with a material, its intensity I decreases exponentially as a function of the sample thickness. This behavior can be expressed in terms of the incident radiation intensity I_0 (associated with the initial radiation intensity before entering the sample) and the transmitted intensity I_T according to the following mathematical expression:

$$I_T = I_0 e^{-\alpha x} \tag{1.42}$$

where x is the thickness of the sample and α is the absorption coefficient.

The penetration depth of electromagnetic radiation in materials is defined as the distance at which the radiation intensity has decayed to a proportion of e^{-1}, that is:

$$d = \frac{1}{\alpha} \tag{1.43}$$

where d is the penetration depth and α is the absorption coefficient. In Table 1.2, the penetration depths of various lasers used in a series of semiconductor samples are shown.

The data in Table 1.2 are very helpful when analyzing the interaction of laser radiation with materials. For example, in characterization techniques such as photoreflectance, photoluminescence, and Raman spectroscopy, the penetration depth of the laser used in the measurement is important for approximately defining the region of the sample from which the optical information originates. For the spectra obtained as part of this work, the wavelength used in the Raman system is 473 nm, and according to Table 1.2, the penetration depth for GaAs samples is approximately 500 nm.

Table 1.2 Penetration depths for the different lasers used in optical measurements are shown

Laser	Energy	Coefficient extinction	Penetration depths	Coefficient extinction	Penetration depths
λ	eV	k_{GaAs}	d_{GaAs}	k_{ZnSe}	d_{ZnSe}
Å	eV		Å		Å
6328	1.959	0.19	2650	—	—
5430	2.283	0.31	1394	—	—
4880	2.541	0.74	525	—	—
4416	2.808	1.46	240	0.16	2200
3250	3.815	2.00	129	0.53	490

Example Photons from a laser with a wavelength of $\lambda = 638$ nm are incident on the surface of a GaAs semiconductor. Given that the extinction coefficient is $k = 0.19$, what is the penetration length of these photons?

$$\left| \begin{array}{l} \lambda_{Laser} = 632.8\,\text{nm} \\ k_{GaAs} = 0.19 \end{array} \right. \rightarrow \alpha = \frac{4 \cdot \pi \cdot k_{GaAs}}{\lambda_{Laser}} \rightarrow \alpha = 3.773 \times 10^4 \frac{1}{\text{cm}} \tag{1.44}$$

Then

$$d = \frac{1}{\alpha} \rightarrow d = 265.04\,\text{nm} \tag{1.45}$$

The penetration depth depends strongly on the wavelength of the incident light. Shorter wavelengths (higher energy photons) are typically absorbed more quickly, resulting in shorter penetration depths. Absorption coefficient α is a specific parameter for materials and quantifies how strongly the semiconductor absorbs light at a given wavelength. It is a function of the materials' electronic structure and the photon energy. Understanding the depth of penetration in materials, especially semiconductors, is essential to design efficient experiments and obtain defined Raman spectra. It allows the researcher to obtain information from specific surface regions or layers within the material.

Optical Density Filter The absorption coefficient α of an optical medium can also be quantified in terms of the parameter called Optical Density (OD). This parameter is very useful for designing optical filters. OD is a physical quantity that measures how much a material absorbs per unit length. The optical materials used for this purpose are generally transparent. The behavior of materials when exposed to light, referred to as the interaction of radiation with matter, strongly depends on the incident wavelength (λ).

Absorbance (A), also known as optical density (OD), is a measure of how much light is absorbed by a sample as it passes through it. It is a key concept in spectrophotometry. An optical density filter (ODF), also known as a neutral density (ND) filter, is a type of filter used in photography and scientific applications to reduce the intensity of light without changing its color. The primary purpose of an ND filter is to allow for more control over exposure settings by reducing the amount of light entering the lens. This is useful in various scenarios, such as long-exposure photography, reducing depth of field in bright conditions, and preventing overexposure. There are different types of ND filters based on their optical density:

Fixed ND filters: These have a set optical density.

Variable ND filters: These allow the user to adjust the optical density by rotating the filter elements.

ND filters are used to achieve motion blur in moving water, clouds, or other subjects by allowing for longer exposure times. They are also used to control depth of field in bright conditions by allowing for wider apertures. Additionally, ND filters are used in various optical experiments to control the intensity of light in a controlled manner without affecting other properties of the light. ND filters are typically made from glass or resin coated with a neutral material that evenly attenuates light across the visible spectrum. High-quality ND filters ensure that the color balance of the image remains unchanged.

Thus, OD is a function of λ. Therefore, it is defined as

$$A(\lambda) = DO(\lambda) \tag{1.46}$$

where $A(\lambda)$ is the optical absorbance of the material, dependent on the wavelength of the incident light. $A(\lambda)$ is defined using the transmittance through the material as

$$A(\lambda) = \log_{10} T = \log_{10} \frac{I_0}{I_T} \tag{1.47}$$

Then, OD can be expressed using the absorbance, according to Eq. 1.46:

$$OD = \log_{10}\left(\frac{I_0}{I_T}\right) \tag{1.48}$$

where I_0 is the intensity of the incident light and I_T is the intensity of the transmitted light. This relationship is known as the Beer-Lambert law (see Eq. 1.42.). Then,

$$\begin{cases} OD(\lambda) = A(\lambda) = \log_{10} \frac{I_0}{I_T} = \log_{10} \frac{I_0}{I_0 e^{-\alpha x}} = \\ = \log_{10}(e^{\alpha x}) = \alpha x \log_{10} e = 0.434\alpha x \end{cases} \tag{1.49}$$

Table 1.3 Optical density

Optical density (OD)	Transmission percent of the laser intensity
0.0	100
0.3	50
0.6	25
1	10
2	1
3	0.1
4	0.01

$$OD(\lambda) = 0.434\alpha x \tag{1.50}$$

In Table 1.3 , the OD of the filter used is related to the proportion by which the intensity of the incident laser on the sample decreases.

For example, when a filter with OD =1.0 is used, it means that the intensity has decreased to $0.1 I_0$:

$$10^{-OD} = \frac{I(l)}{I_0} \rightarrow I(l) = I_0 \cdot 10^{-OD} \tag{1.51}$$

$$OD = 1.0 \rightarrow I(l) = 0.1 I_0 \tag{1.52}$$

Consider a sample with an absorption coefficient $\alpha = 0.5cm^{-1}$, and the path length (x) is 1 cm. The optical density (OD) is

$$OD(\lambda) = 0.434\alpha x = 0.434(0.5\text{cm}^{-1})(2\text{cm}) = 0.217 \tag{1.53}$$

This means that over a 1 cm path, the sample has an optical density of approximately 0.217.

Optical density (OD) is a dimensionless measure of light attenuation over a specific path length. At the same time, the absorption coefficient quantifies the attenuation per unit length and has units of inverse length.

OD: Calculated using the incident to transmitted light intensities ratio over a specific path length.

Absorption Coefficient: Describes the attenuation of light per unit length of the material.

Generally, laser powers around 50 mW are used for ceramic materials and most semiconductors. However, laser powers around 5 mW must be employed in some oxides like manganese oxides and low-melting-point polymers. In such cases, neutral density filters with an OD of around 1.0 are necessary. Table 1.3 shows the filter number and the percentage of light signal reduction.

It is essential to perform preliminary measurements of the incident laser powers on the sample. A good approach is to use powers around 50 mW and focus on a representative peak in the sample's response spectrum. This peak is typically referred to as the control peak. By recording the spectrum around the control peak, one can observe if the shape of the curve changes as the scan is repeated. If this is the case, the power should be reduced until the curve shape remains consistent, ensuring that the signal−noise ratio is optimized.

Another important consideration in the use of lasers in spectroscopic techniques includes Rayleigh scattering processes. Under the assumption that the size of the scattering center is much smaller than the wavelength of the electromagnetic radiation used, the scattering cross section $\sigma_s(\lambda)$ varies as a function of the wavelength λ according to

$$\sigma_s(\lambda) \propto \frac{1}{\lambda^4} \tag{1.54}$$

The intensity of Rayleigh scattering is inversely proportional to the fourth power of the wavelength, meaning shorter wavelengths (blue light) scatter more than longer wavelengths (red light). This scattering is an elastic process, meaning the scattered photons have the same energy (wavelength) as the incident photons. In inhomogeneous materials and most binary oxides, short wavelengths produce significant elastic scattering, which enhances fluorescence and shields the Raman signal. This necessitates increasing the signal collection time or, if available, switching to a laser with longer wavelengths, such as those in the red or near-infrared region.

1.3 Units for Energy

Electromagnetic radiation, from a classical perspective, is considered as coupled perpendicular electric and magnetic fields, while from a quantum perspective, it is visualized as a collection of particles called photons. In optical spectroscopy, it is useful to employ the expression that relates the energy of the photon to the wavelength. In Raman spectroscopy, the intensity of the electromagnetic radiation signal I is generally used as a function of the wavenumber, where the wavenumber

Table 1.4 Factors used in unit conversion

	1 cm^{-1}	1 kJ mol^{-1}	1 eV	1 $kcalmol^{-1}$
1 cm^{-1}	1		1.23985×10^{-4}	2.8591×10^{-3}
1 kJ mol^{-1}		1		
1 eV	8065.5		1	23.060
1 kcal mol^{-1}	349.75		4.3364×10^{-2}	1

is defined as the reciprocal of the wavelength $\tilde{\nu}$. It has been standardized that the units of $\tilde{\nu}$ are cm^{-1}.

$$\tilde{\nu} = \frac{1}{\lambda} \tag{1.55}$$

In calculations, the following useful conversions are employed:

$$\nu\left[cm^{-1}\right] = \frac{10^4}{\lambda(\mu m)}, \quad \nu\left[cm^{-1}\right] = \frac{10^7}{\lambda(nm)} \tag{1.56}$$

The conversion factor used to relate the respective units is shown in Table 1.4.

1.3.1 *Wavenumber*

Wavenumber is a measure used in spectroscopy, defined as the number of wavelengths per unit distance, and is commonly expressed in reciprocal centimeters (cm^{-1}). For example, a wavenumber of 1000 cm^{-1} means that there are 1000 wavelengths in one centimeter of space.

The wavenumber $\tilde{\upsilon}$ is defined as the reciprocal of the wavelength λ. If the wavelength λ is expressed in centimeters, the wavenumber will be in cm^{-1}. The wavenumber is directly proportional to the frequency f of the wave and, therefore, to the energy E of the photon associated with the wave. The relationship to frequency is given by

$$E = hf = \frac{hc}{\lambda} = hc\tilde{\upsilon} \tag{1.57}$$

where

$$\frac{1}{\lambda} = \tilde{\upsilon} \tag{1.58}$$

and h and c are Planck's constant and the speed of light, respectively.

$$\left| \begin{array}{l} h = 6.626 \times 10^{-34}\ J \cdot s \\ c = 2.998 \times 10^{8}\ \frac{m}{s} \end{array} \right. \tag{1.59}$$

In spectroscopy, wavenumbers are often used instead of wavelengths because they provide a more direct measure of energy levels and transitions in molecules. For example, in infrared (IR) spectroscopy and Raman spectroscopy, wavenumbers are used to describe vibrational transitions in molecules. Wavenumbers are commonly used because they are directly proportional to energy, making them a natural unit for comparing different transitions or spectral features.

Example, what is 1 eV in cm^{-1}?

For E = 1 eV, we obtain

$$E = 1\ eV \rightarrow \tilde{\upsilon} = \frac{E}{hc} = 8.065 \times 10^{3}\ \mathrm{cm}^{-1} \tag{1.60}$$

Example
Silicon exhibits a strong and sharp Raman peak at around $\tilde{\upsilon} = 520.7\mathrm{cm}^{-1}$, corresponding to the optical phonon mode in its crystal structure. What is the energy of this optical phonon in eV?

$$E = hc\tilde{\upsilon} = \left(6.626 \times 10^{-34}\ J \cdot s\right)\left(2.998 \times 10^{8}\ \frac{m}{s}\right)\left(520.7\ \mathrm{cm}^{-1}\right) = 65\ \mathrm{meV} \tag{1.61}$$

1.4 Estimation of Rotational and Vibrational Parameters

The study of any stationary physical system, such as solids, molecules, or atoms, involves solving the time-independent Schrödinger equation:

$$\hat{H}\Psi = E\Psi \tag{1.62}$$

where $\hat{H}$ is the Hamiltonian operator, Ψ is the wave function of the system, and E is the energy eigenvalue associated with $\hat{H}$. This fundamental equation describes how the quantum state of a physical system evolves and is central to quantum mechanics. The Hamiltonian operator $\hat{H}$ is written as

$$\hat{H} = \frac{\hbar^2}{2m}\nabla^2 + V \tag{1.63}$$

where ∇^2 and $V(x, y, z)$ are the Laplacian operator and potential energy function, respectively. In Cartesian coordinates, the time-independent Schrödinger equation is expressed as

$$-\frac{\hbar^2}{2m}\nabla^2\Psi(x, y, z) + V(x, y, z)\Psi = E\Psi \tag{1.64}$$

and

$$\frac{\hbar^2}{2m}\left(\frac{\partial^2\psi(r)}{\partial x^2} + \frac{\partial^2\psi(r)}{\partial y^2} + \frac{\partial^2\psi(r)}{\partial z^2}\right) + U(r)\psi(r) = E\psi(r) \tag{1.65}$$

Therefore, the Schrödinger equation written in the time-independent operator form is

$$\hat{H}\psi(r) = E\psi(r) \tag{1.66}$$

For the time-dependent part, we obtain

$$-i\hbar\left(\frac{\partial^2 f(t)}{\partial t^2}\right) = Ef(t) \tag{1.67}$$

and

$$Ef(t) = e^{\pm\frac{iEt}{\hbar}} \tag{1.68}$$

The equation for the complete wave function is

$$\Psi(r, t) = \psi(r)e^{\pm\frac{iEt}{\hbar}} \tag{1.69}$$

where E_n is determined by spectroscopic techniques and $\hat{H}$ is the so-called Hamiltonian operator.

harmonic oscillator When the Schrödinger equation is solved for a **harmonic oscillator**, the obtained energy values are

$$E_n = (n + \frac{1}{2})\hbar\omega \tag{1.70}$$

and n = 0, 1, 2....

$$\Delta E = \hbar\omega \tag{1.71}$$

When the Schrödinger equation is solved for the **many-electron atom model**, the energy levels are obtained through

$$E_n = \frac{-13.6Z_{efect}^2}{n^2} eV \tag{1.72}$$

with n = 0, 1, 2...

Example: A harmonic oscillator undergoes an electronic transition from the n = 1 to the n = 3 state. What is the energy required to produce this transition?

To determine the energy required for a transition in a quantum harmonic oscillator, we use the formula for the energy levels of a harmonic oscillator, which is given by

$$E_n = \left(n + \frac{1}{2}\right) \hbar\omega \tag{1.73}$$

where:

- E_n is the energy of the n-th level,
- $\hbar$ is the reduced Planck's constant, and
- ω is the angular frequency of the oscillator.

To find the energy required for a transition from the n = 1 to the n = 3 state, we need to calculate the energy difference between these two states:

Energy of the $n = 1$ **state**:

$$E_1 = \left(1 + \frac{1}{2}\right) \hbar\omega = \frac{3}{2}\hbar\omega$$

Energy of the $n = 3$ **state**:

$$E_3 = \left(3 + \frac{1}{2}\right) \hbar\omega = \frac{7}{2}\hbar\omega$$

The energy required for the transition from $n = 1$ to $n = 3$ is the difference between these two energy levels:

$$\Delta E_{1-3} = E_3 - E_1 = \frac{7}{2}\hbar\omega - \frac{3}{2}\hbar\omega = 2\hbar\omega$$

Therefore, the energy required to produce the transition from the $n = 1$ state to $n = 3$ state in a harmonic oscillator is $2\hbar\omega$.

The wave function for atoms, Ψ, is characterized by five quantum numbers: n, l, m_l, s, m_s, which have the properties listed in Table 1.5, and are compatible with the selection rules that allow or define the permitted transitions between different states. These rules are governed by

$$\Delta l = \pm l \tag{1.74}$$

Table 1.5 Quantum numbers

Quantum number	Symbol	Orbital	Value range	Example
Principal	n	Shells	$n \geq 1$	$n = 1, 2...$
Angular or orbital momentum	l	Subshells	$0 \leq l \leq n - 1$	$n = 3, l = 0, 1, 2(s, p, d)$
Magnetic	m_l	Orbital	$-l \leq m_l \leq l$	$l = 2, m_l = -2, -1, 0, 1, 2$
Spin	S	Spin	$\frac{1}{2}$	
Spin projection	m_s	Spin	$-\frac{1}{2}, \frac{1}{2}$	$-\frac{1}{2}, \frac{1}{2}$

$$\Delta m_l = 0, \pm 1 \tag{1.75}$$

Electronic transitions in a quantum system occur from an initial level with energy E_i to a final level with energy E_f, respecting the selection rules. The perturbation that causes the transition in optical spectroscopy techniques can be due to incident electrons on the atom or photons. The process is illustrated in Fig. 1.1.

In quantum mechanics, angular momentum is a fundamental quantity related to the rotational symmetry of molecules in a system. It is important due to its role in understanding the behavior of particles at the quantum level. There are two types of angular momentum in quantum mechanics: orbital angular momentum and spin angular momentum.

Orbital Angular Momentum: Orbital angular momentum arises from the motion of a particle around a central point (e.g., an electron orbiting a nucleus). It is quantized, meaning it can only take on discrete values. The quantum numbers associated with orbital angular momentum are

- Principal Quantum Number (n): Determines the energy level of the electron.
- Orbital Quantum Number (l): Determines the shape of the orbital and can take integer values from 0 to n.

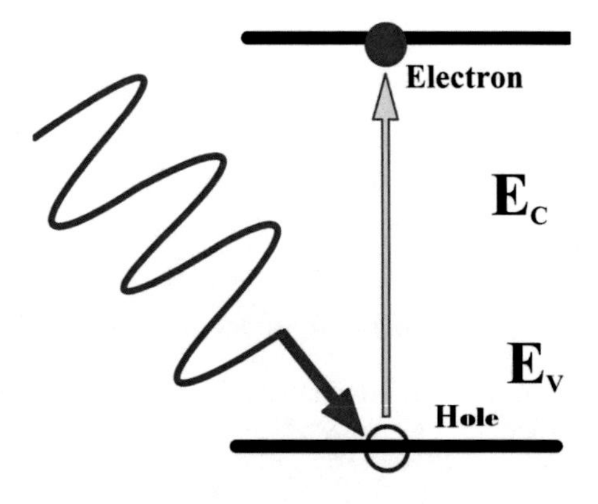

Fig. 1.1 Model for the photon absorption

- Magnetic Quantum Number (ml): Determines the orientation of the orbital in space and can take integer values from $-l$ to $+l$.

The magnitude of the orbital angular momentum is given by

$$\mathbf{L} = \sqrt{l(l+1)}\hbar \tag{1.76}$$

where $\hbar$ is the reduced Planck's constant.

Spin Angular Momentum: Spin angular momentum is an intrinsic form of angular momentum possessed by particles, independent of any motion through space. It is also quantized and described by the spin quantum number (S) and the spin magnetic quantum number (m_s).
where

- **Spin Quantum Number**: For particles as electrons, the spin takes a value $s = \frac{1}{2}$.
- **Spin Magnetic quantum Number**: For particles as electron, the spin magnetic or spin projection takes a value $m_s = \frac{1}{2}$.

And the magnitude of the spin angular momentum is given by

$$|\mathbf{S}| = S = \sqrt{s(s+1)}\hbar \tag{1.77}$$

Total Angular Momentum Total angular momentum is defined as **J**, and this is written as the vector sum of the orbital **L** and spin angular momentum **S**.

$$\mathbf{J} = \mathbf{L} + \mathbf{S} \tag{1.78}$$

and J takes the values given below:

$$j = l + s,\ l + s - 1,\ \ldots, |l - s| \tag{1.79}$$

For a diatomic molecule, the rotational energy levels can be described by quantum mechanics. The energy levels are quantized and given by

$$E_J = \frac{\hbar^2}{2I} J(J+1) \tag{1.80}$$

where:

- E_J is the rotational energy for the rotational quantum number J;
- $\hbar$ is the reduced Planck constant ($\hbar = \frac{h}{2\pi}$);
- I is the moment of inertia of the molecule;
- J is the rotational quantum number $J = 0, 1, 2...$

Moment of Inertia The moment of inertia I concerning mass center of a diatomic molecule is given by

has been defined that

$$\mathbf{r} = x\hat{i} \tag{1.87}$$

$$I = \int r^2\, dm \Rightarrow \sum_0 r_i^2 m_i = \frac{x^2}{4} m_1 + \frac{x^2}{4} m_2 \tag{1.88}$$

$$x = (x_1 - x_2) - L \tag{1.89}$$

is the bond length or the length of the imaginary spring when the diatomic e model consists of a spring with constant and two masses m_1 and m_2, as n Fig. 1.2.

$$m_1 \frac{\partial^2 x_1}{\partial t^2} = -kx \tag{1.90}$$

$$m_2 \frac{\partial^2 x_2}{\partial t^2} = +kx \tag{1.91}$$

$$m_1 m_2 \frac{\partial^2 x_1}{\partial t^2} + m_1 m_2 \frac{\partial^2 x_2}{\partial t^2} = -(m_1 + m_2)kx \tag{1.92}$$

$$m_1 m_2 \frac{\partial^2 (x_1 - x_2)}{\partial t^2} = -(m_1 + m_2)kx \tag{1.93}$$

$$\frac{m_1 m_2}{(m_1 + m_2)} \frac{\partial^2 (x_1 - x_2 - L)}{\partial t^2} = -kx \tag{1.94}$$

$$\mu = \frac{m_1 m_2}{(m_1 + m_2)} \tag{1.95}$$

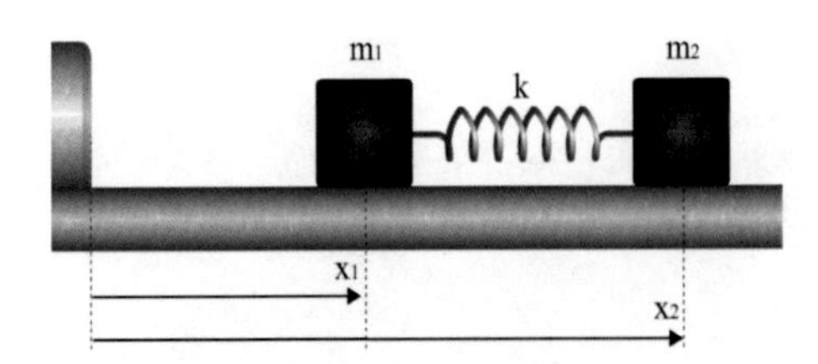

.2 Model for molecule nic

$$I = \mu r^2$$

with

$$\mu = \frac{m_1 m_2}{m_1 + m_2}$$

Selection Rules
In rotational spectroscopy, transitions between rotationa
to the absorption or emission of photons. According to
selection rules for these transitions are

$$\Delta J = \pm 1$$

Equation indicates that a molecule can undergo a transiti
with quantum number J to a state with quantum number J

1.4.1 Rotational Energy of Molecules

When molecules compress or expand around the equilibriu
as vibrational motion. In the case where molecules rotate arc
we are observing rotational motion. Materials are compose
that consist of nuclei and electrons, and from an energetic
electronic ($E_{\text{electronic}}$), rotational ($E_{\text{rotational}}$), vibrational (E_{vibra}
($E_{\text{translational}}$) level energies. The total energy can be describe

$$E_T = E_{electronic} + E_{translational} + E_{rotational} + E$$

Diatomic Molecule: Rotational energy in molecules is the e
the rotational movement of the molecule around its center of
fundamental in the study of molecular spectroscopy and the ap
mechanics. In the case of a diatomic molecule, it is possible t
as shown in Fig. 1.2, where two bodies of different masses (m_1
spring with constant K are considered.

With,

$$E_{rotational} = \frac{1}{2} I \omega^2$$

ω is angular frequency and I is the so-called moment of inertia
for a diatomic molecule by

$$I = (\frac{m_1 m_2}{m_1 + m_2}) r^2 = \mu r^2$$

$$\mu \frac{\partial^2 x}{\partial t^2} = -kx \tag{1.96}$$

$$\frac{\partial^2 x}{\partial t^2} + \frac{k}{\mu} x = 0 \tag{1.97}$$

The expression (1.97) is a wave equation of the type:

$$\ddot{u} + \omega^2 u = 0 \tag{1.98}$$

whose solution is

$$u = u_m (e^{i\omega t} - e^{-i\omega t}) \tag{1.99}$$

where u_m is the maximum amplitude of the oscillatory motion. By comparing Eqs. 1.97 and 1.98, we infer that

$$\omega = \sqrt{\frac{k}{m}} \tag{1.100}$$

The constant k is related with bond strength.

$$E_{Rot} = \frac{1}{2} I \omega^2 = \frac{1}{2} (\frac{m_1 m_2}{m_1 + m_2}) r^2 \omega^2 \tag{1.101}$$

$$\mathbf{L} = I\omega = \sqrt{J(J+1)}\hbar \tag{1.102}$$

with J = 0, 1, 2 ... and as

$$J = L + S \tag{1.103}$$

Then,

$$E = \frac{1}{2} I \omega^2 = \frac{1}{2I} (I\omega)^2 = \frac{1}{2I} (J(J+1)\hbar^2 \Rightarrow E_{rot} = \frac{\hbar^2}{2I} J(J+1) \tag{1.104}$$

as

$$\Delta E = E_f - E_i \tag{1.105}$$

So that

$$\Delta E = E_{J+1} - E_J \Rightarrow \Delta E = \frac{\hbar^2}{2I}\{(J+1)[(J+1)+1] - J(J+1)\} \quad (1.106)$$

$$\Delta E = \frac{\hbar^2}{2I}[(J+1)(J+2) - J(J+1)] \quad (1.107)$$

$$\Delta E = \frac{\hbar^2}{2I}[(J+1)(J+2-J)] \quad (1.108)$$

$$\Delta E = \frac{\hbar^2}{2I}[2(J+1)] \quad (1.109)$$

$$\Delta E = \frac{\hbar^2}{I}(J+1) \Rightarrow \Delta E \approx \frac{h^2}{4\pi I}J = h\nu \quad (1.110)$$

Example applied to the CO molecule

By solving the problem step by step and considering that the transition frequency from J = 0 to J = 1 occurs at

$$\nu = 1,1510^{11} Hz \quad (1.111)$$

The rotational transition energy is calculated as

$$\Delta E \approx \frac{h^2}{4\pi I}J = \frac{h^2}{4\pi I} = h\nu \quad (1.112)$$

The moment of inertia I obtained is

$$I = 1,46 \times 10^{-46}\text{Kg} * \text{m}^2 \quad (1.113)$$

The reduced mass is calculated using the atomic masses of carbon (^{12}C) and oxygen (^{16}O). The atomic mass in atomic mass units (μ) is

$$\mu = (\frac{m_1 m_2}{m_1 + m_2}) = \frac{(12u)(16u)}{12u + 16u} = 1,14 \times 10^{-26}\text{kg} \quad (1.114)$$

where

$$u = 1,66 \times 10^{-27}\text{kg} \quad (1.115)$$

The bond length of the CO molecule is determined using the relationship between the moment of inertia and reduced mass:

$$r = \sqrt{\frac{I}{\mu}} = 1,13 \times 10^{-10}m = 1,13\mathring{A} \tag{1.116}$$

Therefore, the bond length of the CO molecule is approximately $\mathring{A}$.

Example applied to Hydrogen Chloride (HCl)

In the following example, with other information available, we can perform some calculations. Consider the diatomic molecule HCl. The rotational energy levels for HCl can be calculated using the above formulas. The bond length r for HCl is approximately 1.27Å (angstroms), and the masses of hydrogen and chlorine are approximately 1.6710^{-27} and 5.8110^{-26} kg, respectively.

The reduced mass μ is

$$\mu = \frac{1.67 \times 10^{-27}\text{kg} \times 5.81 \times 10^{-26}\text{kg}}{1.67 \times 10^{-27}\text{kg} + 5.81 \times 10^{-26}\text{kg}} \approx 1.62 \times 10^{-27}\ \text{kg} \tag{1.117}$$

The moment of inertia I is

$$I_{HC} = 1.62 \times 10^{-27}\text{kg} \times \left(1.27 \times 10^{-10}m\right)^2 = 2.62 \times 10^{-47}\text{kgm}^2 \tag{1.118}$$

The rotational energy levels E_J are given by

$$E_J = \frac{\hbar^2}{2I}J(J+1) \tag{1.119}$$

and for selection rule $\Delta E = \frac{\hbar^2}{I}(J+1)$, and J=1, then

$$\Delta E = \frac{h^2}{4\pi I}2 = \frac{h^2}{2\pi I} \tag{1.120}$$

Then

$$\Delta E = \frac{\left(6.626 \times 10^{-34}J \cdot s\right)^2}{2\pi \times 2.62 \times 10^{-47}\text{kg} \cdot m^2} = 2.67 \times 10^{-21}J \tag{1.121}$$

The rotational spectrum of a molecule like HCl will consist of lines corresponding to transitions between these quantized energy levels. Each line in the spectrum corresponds to a photon absorbed or emitted during a transition, and the frequency of the photon is related to the energy difference between the initial and final states.

1.4.2 Vibrational Energy of the Molecules

The model of a mass m attached to a spring with constant k applies to the case of a system of two particles with masses m_1 y m_2 (see Fig. 1.2). In this case, the mass in the model is replaced by the reduced mass, so that the equation is expressed as

$$\frac{\partial^2 x}{\partial t^2} + \frac{k}{\mu}x = 0 \tag{1.122}$$

This differential equation is second order, linear, and has constant coefficients, whose solution is

$$X = Ae^{-i\omega_0 t} \tag{1.123}$$

Taking the first and second derivatives with respect to time of expression (1.123), we obtain

$$X' = -iw_o Ae^{-i\omega_0 t} \tag{1.124}$$

$$X'' = -w_o^2 Ae^{-i\omega_0 t} \tag{1.125}$$

By substituting the previous expressions into Eq. 1.122, it yields

$$-w_o^2 Ae^{-i\omega_0 t} + \frac{k}{\mu}Ae^{-i\omega_0 t} = 0 \tag{1.126}$$

where w_o^2 is defined as the natural or characteristic frequency of the system.

$$w_o^2 = \frac{k}{\mu} \tag{1.127}$$

$$w_o = \sqrt{\frac{k}{\mu}} \tag{1.128}$$

Moreover,

$$w_o = \frac{2\pi}{T} = 2\pi v \tag{1.129}$$

Therefore, the natural frequency of the system is

$$v = \frac{1}{2\pi}\sqrt{\frac{k}{\mu}} \tag{1.130}$$

The quantum energy for a harmonic oscillator is obtained from the solution of the Schrödinger equation, and its expression is

$$E_{vibracion} = \left(n + \frac{1}{2}\right)\hbar\omega = \left(n + \frac{1}{2}\right)\hbar\sqrt{\frac{k}{\mu}} \tag{1.131}$$

We substitute the classical expression for the frequency, given by Eq. 1.130, into Eq. 1.131 to obtain

$$E_{vibracion} = \left(n + \frac{1}{2}\right)\hbar\sqrt{\frac{k}{\mu}} \tag{1.132}$$

The selection rule for the quantum number n is

$$\Delta n = \pm 1 \tag{1.133}$$

Therefore, the energy difference between consecutive levels allowed by the selection rule is

$$\Delta E_{vib} = E_n - E_{n-1} \tag{1.134}$$

Thus:

$$\left(\left(n + \frac{1}{2}\right) - \left(n + \frac{1}{2}\right)\right)\hbar\sqrt{\frac{k}{\mu}} = \hbar\sqrt{\frac{k}{\mu}} \tag{1.135}$$

Finally, the vibrational energy for the diatomic molecule with masses m_1 and m_2 is

$$E_{vib} = \hbar\sqrt{\frac{k}{\mu}} \tag{1.136}$$

It is assumed for the calculations that the vibration is around the equilibrium position. For the vibration of the CO molecule in the transition from n = 0 to n = 1, it is experimentally observed that the transition occurs at

$$\nu = 6,42x10^{13}Hz \tag{1.137}$$

The value of k is determined using the following relation:

$$\hbar\sqrt{\frac{k}{\mu}} = \hbar\omega = \hbar 2\pi\upsilon \tag{1.138}$$

$$k = 4\pi^2 \mu v^2 \tag{1.139}$$

where

- $\hbar$ is the reduced Planck's constant ($\hbar = \frac{h}{2\pi}$) and
- μ is the reduced mass of the CO molecule.

By substituting the values $\mu = 1.14 \times 10^{-26}$ and $\nu = 6, 42 \times 10^{13}$ Hz, we obtain the value of k:

$$k = 1.85x10^3 \frac{N}{m} \tag{1.140}$$

Calculation of the Maximum Amplitude of Vibration The maximum amplitude of vibration A is determined using the potential energy at the maximum position:

$$E_{pot} = \frac{1}{2} K A^2 = \hbar \sqrt{\frac{k}{\mu}} \tag{1.141}$$

From which, we obtain

$$A = 4, 79x10^{-3} nm \rightarrow A = 4, 79x10^{-2} \mathring{A} \tag{1.142}$$

Comparing the maximum amplitude of the vibration A obtained in Eq. 1.142 with the bond length r obtained previously in (1.116), it is observed that the vibrations are two orders of magnitude smaller than the bond length. This confirms that the vibration is relatively small compared to the bond length, which is consistent with the assumption that the vibrations occur around the equilibrium position.

1.5 Proposed Problems

1.1 Calculate the refractive index of BK7 glass at two specific wavelengths ($\lambda = 0.55\mu$m and $\lambda = 0.63\mu$m). Suppose that the Sellmeier coefficients B_i and C_i are the same, and they are reported with the following values:

- $B_1 = 1.03961212$
- $B_2 = 0.231792344$
- $B_3 = 1.01046945$
- $C_1 = (0.00600069867)^2$
- $C_2 = (0.0200179144)^2$
- $C_3 = (103.560653)^2$

1.2 A semiconductor material has a bandgap $E_g = 1.49\text{eV}$. A laser with a wavelength of 783 nm is incident on the surface of this material. First, indicate if the semiconductor can absorb this light and, if so, calculate the extinction coefficient k when the material has an the absorption coefficient of $\alpha = 1.5 \times 10^{-4}\text{cm}^{-1}$ at this wavelength.

1.3 A semiconductor sample of thickness 1 μm shows a decrease in light intensity from 50 $\frac{mW}{\text{cm}^2}$ to 5 $\frac{mW}{\text{cm}^2}$ at a wavelength of 632.8 nm. Calculate the absorption coefficient α.

1.4 A semiconductor film with a thickness of 0.5 μm has an optical density (OD) of 3 at a specific wavelength. Calculate the absorption coefficient α.

1.5 A commercial filter allows one-quarter of the incident light intensity to pass through. What is the optical density of this filter?

1.6 Consider a sample with an extinction coefficient ($\alpha = 0.4\text{cm}^{-1}$, and the path length (x) is 3 cm. What is the optical density (OD)?

1.7 A harmonic oscillator undergoes an electronic transition from the $n = 2$ state to the $n = 5$ state. What is the energy required to produce this transition?

1.8 What is the incident photon energy required to produce an electronic transition from $n = 2$ state to $n = 5$ state?

1.9 A linear diatomic molecule (like CO) has a vibrational transition with a wavenumber of 2143 cm^{-1}. What is the corresponding energy in joules?

1.10 In ZnO, the modes at 332 and 380 cm^{-1} are due to multi-phonon processes, indicating better crystalline quality of the samples. What are the energies of these peaks in eV?

References

1. Serway RA, Jewett JW (2018) Physics for scientists and engineers with modern physics, 10th ed. Cengage Learning
2. Cheng DK (1993) Fundamentals of engineering electromagnetics. Addison-Wesley
3. Cohen-Tannoudji C, Diu B, Laloë F (1977) Quantum mechanics, vols 1 and 2. Wiley
4. Pedrotti FL, Pedrotti L (2017) Introduction to optics, 3rd ed. Pearson
5. Dirac PAM (1981) The principles of quantum mechanics, 4th ed. Oxford University Press
6. Jackson JD (1998) Classical electrodynamics, 3rd ed. Wiley
7. Griffiths DJ (2017) Introduction to electrodynamics, 4th ed. Cambridge University Press
8. Vargas Hernández C (2005) Interacciones fotónicas en películas semiconductoras y su caracterización. Universidad Nacional de Colombia. ISBN 958-9322-96-4

Chapter 2
Quantum Mechanics Theory

In this text, we will occasionally use the notation introduced by Paul Dirac, which is associated with the dot product between functions. The parentheses or brackets in English allow us to define the following [1–3]:

$$\left|\begin{array}{l}\langle\Psi \mid \Psi\rangle \rightarrow \text{Bracket} \\ \langle\Psi| \rightarrow \text{Bra} \\ |\Psi\rangle \rightarrow \text{Ket}\end{array}\right. \tag{2.1}$$

Using this notation, we can define the expected (or mean) value of a physical observable associated with the operator $\hat{\mathrm{A}}$:

$$\langle \mathrm{A}\rangle = \int_{-\infty}^{\infty} \psi_m^* \hat{\mathrm{A}} \psi_n d\tau \equiv \langle \psi_m | \hat{\mathrm{A}} | \psi_n \rangle \equiv \langle m | \hat{\mathrm{A}} | n \rangle \equiv \hat{\mathrm{A}}_{mn} \tag{2.2}$$

In solving these systems, the state functions $\psi(r, t)$ depend on parameters or integers, similar to the state functions found for a one-dimensional infinite potential box, which are described as follows:

$$\left|\begin{array}{l}\psi_n(x) = \left(\frac{2}{L}\right)^{1/2} sen\left(\frac{n\pi x}{L}\right) \quad 0 < x < L \\ \\ \psi_n(x) = 0 \quad \text{otherwise}\end{array}\right. \tag{2.3}$$

Thus, its specification is determined by the parameter n. It should be noted that the first letter of the bracket indicates that the function is conjugate. Using the brackets, we can define the integral of the product of two functions as follows:

C. Vargas Hernandez, *Introduction to Raman Spectroscopy and Its Applications*,
https://doi.org/10.1007/978-3-031-77551-2_2

$$\int_{-\infty}^{\infty} \psi_m^* \psi_n d\tau = \langle m \mid n \rangle, (AB)^* = B^* A^* \tag{2.4}$$

Another important relationship is the conjugate of the multiplication of operators, such as

$$(AB)^* = B^* A^* \tag{2.5}$$

Applying the above equation, we have

$$\left(\int_{-\infty}^{\infty} \psi_m^* \psi_n d\tau \right)^* = (\langle m \mid n \rangle)^* = \int_{-\infty}^{\infty} \psi_n^* \psi_m d\tau = \langle n \mid m \rangle \tag{2.6}$$

Therefore, we have

$$\langle m \mid n \rangle^* = \langle n \mid m \rangle \tag{2.7}$$

2.1 Postulates of Quantum Mechanics

2.1.1 Postulate 1

The state of a system is described by the function $\psi(r, t)$, which contains all the information about the system.

The state function $\psi(r, t)$ has the following requirements:

1. $\psi(r, t)$ It is single valued.
2. $\psi(r, t)$ It is continuous.
3. The derivatives of $\psi(r, t)$ are continuous.
4. $\psi(r, t)$ is quadratically integrable, i.e., $\int_{-\infty}^{\infty} \psi^* \psi d\tau$ is finite.

2.1.2 Postulate 2

Every physical observable is associated with a linear Hermitian operator.

Some of the operators used in quantum mechanics are

$$\left| \begin{array}{l} \hat{x} \rightarrow x \\ \hat{E} \rightarrow i\hbar \frac{\partial}{\partial t} \\ \hat{P} \rightarrow -i\hbar \frac{\partial}{\partial x} \end{array} \right. \tag{2.8}$$

2.2 Hermitian Operators

An operator Â is Hermitian if the following is true:

$$\int_{-\infty}^{\infty} f_m^* \hat{A} f_n d\tau = \int_{-\infty}^{\infty} f_n \left(\hat{A} f_m\right)^* d\tau \tag{2.9}$$

Let us explore why quantum mechanics requires Hermitian operators. Let the operator A be associated with a physical observable A. To determine the physical observable A, we perform a measurement on the system, i.e., we find the most probable value of the measurement, which is determined by

$$\langle A \rangle = \int_{-\infty}^{\infty} \psi^* \hat{A} \psi d\tau \tag{2.10}$$

Here, $\langle A \rangle$ is the expected value of the observable A associated with the operator Â. Since the expected value of a physical observable is a real number, we have that

$$\langle A \rangle = \langle A \rangle^* \tag{2.11}$$

$$\langle A \rangle^* = \left(\int_{-\infty}^{\infty} \psi_m^* \hat{A} \psi_n d\tau \right)^* = \int_{-\infty}^{\infty} \left(\hat{A}\psi_n\right)^* \psi_m d\tau = \int_{-\infty}^{\infty} \psi_m \left(\hat{A}\psi_n\right)^* d\tau \tag{2.12}$$

$$\langle A \rangle^* = \left(\langle \psi_m | \hat{A} | \psi_n \rangle \right)^* = \left\langle \hat{A}\psi_n \,\middle|\, \psi_m \right\rangle = \langle \psi_n | \hat{A}^* | \psi_m \rangle = \langle A \rangle = \langle \psi_m | \hat{A} | \psi_n \rangle \tag{2.13}$$

Another way of expressing the Hermitian operator is by noting that

$$\int_{-\infty}^{\infty} f_m^* \hat{A} f_n d\tau = \int_{-\infty}^{\infty} f_n \left(\hat{A} f_m\right)^* d\tau \tag{2.14}$$

is

$$\langle \psi_m | \hat{A} | \psi_n \rangle = \langle \psi_n | \hat{A} | \psi_m \rangle^* \tag{2.15}$$

Additionally, in the previously established notation, it can be written as

$$\hat{A}_{mn} = \hat{A}_{nm}^* \tag{2.16}$$

Let's look at some exercises on the Hermiticity of certain operators. It will start by proving whether or not the operator $\hat{x}$ is a Hermitian. To do this, we must demonstrate that

$$\langle x\rangle = \langle x\rangle^* \tag{2.17}$$

$$\int_{-\infty}^{\infty} f_m^* \hat{x} f_n d\tau = \int_{-\infty}^{\infty} f_n \left(\hat{x} f_m\right)^* d\tau \tag{2.18}$$

$$\int_{-\infty}^{\infty} f_m^* \hat{x} f_n d\tau = \int_{-\infty}^{\infty} f_m^* \hat{x}^* f_n d\tau = \int_{-\infty}^{\infty} f_n \left(\hat{x} f_m\right)^* d\tau \tag{2.19}$$

Next, it will prove that the linear momentum operator $\hat{P}$ is Hermitian. Since it is known that $\hat{P} \to -i\hbar\frac{\partial}{\partial x}$, we must prove that

$$\int_{-\infty}^{\infty} f_m^* \hat{P} f_n d\tau = \int_{-\infty}^{\infty} f_n \left(\hat{P} f_m\right)^* d\tau \tag{2.20}$$

let's see

$$\int_{-\infty}^{\infty} f_m^* \hat{P} f_n d\tau = -i\hbar \int_{-\infty}^{\infty} f_m^* \frac{\partial}{\partial x} f_n dx \tag{2.21}$$

Integrating by parts

$$\int_{-\infty}^{\infty} u dv = uv|_{-\infty}^{\infty} - \int_{-\infty}^{\infty} v du \tag{2.22}$$

then

$$\left| \begin{array}{ll} u(x) = i\hbar f_m^*(x) & dv = \frac{\partial f_n(x)}{\partial x} dx \\ \text{and} & \\ du = i\hbar \frac{\partial f_m^*(x)}{\partial x} dx & v(x) = f_n(x) \end{array} \right. \tag{2.23}$$

$$\int_{-\infty}^{\infty} f_m^* \hat{P} f_n d\tau = -i\hbar \int_{-\infty}^{\infty} f_m^* \frac{\partial}{\partial x} f_n dx = -\, i\hbar f_m^*(x) f_n(x)|_{-\infty}^{\infty} + i\hbar \int_{-\infty}^{\infty} f_n(x) \frac{\partial f_m^*(x)}{\partial x} dx \tag{2.24}$$

Since the functions must be finite as $x \to \pm\infty$, the term uv vanishes. Therefore,

$$\int_{-\infty}^{\infty} f_m^* \hat{P} f_n d\tau = \int_{-\infty}^{\infty} f_n(x) i\hbar \frac{\partial}{\partial x} f_m^*(x) dx = \int_{-\infty}^{\infty} f_n \hat{P}^* f_m^* d\tau = \int_{-\infty}^{\infty} f_n \left(\hat{P} f_m\right)^* d\tau \tag{2.25}$$

$$\int_{-\infty}^{\infty} f_m^* \hat{P} f_n d\tau = \int_{-\infty}^{\infty} f_n \left(\hat{P} f_m\right)^* d\tau \tag{2.26}$$

This indicates that the operator $\hat{P}$ is Hermitian. In bracket notation, we have

$$\int_{-\infty}^{\infty} \psi_m^* \hat{P} \psi_n d\tau = \int_{-\infty}^{\infty} \psi_n \left(\hat{P} \psi_m\right)^* d\tau \tag{2.27}$$

$$\langle \psi_m | \hat{P} | \psi_n \rangle = \langle \psi_n | \hat{P} | \psi_m \rangle^*, \quad \hat{P}_{mn} = \hat{P}_{nm}^* \tag{2.28}$$

2.3 Theorems Associated with Hermitian Operators

2.3.1 T1

When an operator Â is associated with a physical observable A, and it acts on the state function ψ through the eigenvalue equation:

$$\hat{\mathrm{A}} \psi_n = a_n \psi_n \tag{2.29}$$

where the eigenvalue a_n corresponds to a measurable physical quantity, this eigenvalue a_n must be a real number.

Proof Since Â is Hermitian, we have

$$\int_{-\infty}^{\infty} \psi_m^* \hat{\mathrm{A}} \psi_n d\tau = \int_{-\infty}^{\infty} \psi_n \left(\hat{\mathrm{A}} \psi_m\right)^* d\tau \tag{2.30}$$

$$\langle \psi_m | \hat{\mathrm{A}} | \psi_n \rangle = \langle \psi_n | \hat{\mathrm{A}} | \psi_m \rangle^* \tag{2.31}$$

$$\int_{-\infty}^{\infty} \psi_m^* \hat{\mathrm{A}} \psi_n d\tau = \int_{-\infty}^{\infty} \psi_m^* a_n \psi_n d\tau = a_n \int_{-\infty}^{\infty} \psi_m^* \psi_n d\tau = a_n \int_{-\infty}^{\infty} |\psi_{mn}|^2 d\tau = a_n \delta_{mn} \tag{2.32}$$

$$\int_{-\infty}^{\infty} \psi_n \left(\hat{\mathrm{A}} \psi_m\right)^* d\tau = \int_{-\infty}^{\infty} \psi_n \left(a_m \psi_m\right)^* d\tau = \int_{-\infty}^{\infty} \psi_n a_m^* \psi_m^* d\tau = a_m^* \int_{-\infty}^{\infty} |\psi_{mn}|^2 d\tau = a_m^* \delta_{mn} \tag{2.33}$$

$$\left(a_n - a_m^*\right) \delta_{mn} = 0 \rightarrow a_n = a_n^* \tag{2.34}$$

2.3.2 T2

The eigenvalues of a Hermitian operator are real numbers.

Proof

$$\langle m|\,\hat{\text{A}}\,|n\rangle = \langle n|\,\hat{\text{A}}\,|m\rangle^* \tag{2.35}$$

Since,

$$\hat{\text{A}}\,|n\rangle = n\,|n\rangle\,, \quad \hat{\text{A}}\,|m\rangle = m\,|m\rangle \tag{2.36}$$

$$\langle m|\,\hat{\text{A}}\,|n\rangle = \langle m|\,a_n\,|n\rangle = a_n\,\langle m \mid n\rangle = a_n\delta_{mn} \tag{2.37}$$

$$\langle n|\,\hat{\text{A}}\,|m\rangle^* = \langle n|\,a_m\,|m\rangle^* = a_m^*\,\langle n \mid m\rangle^* = a_m^*\delta_{mn} \tag{2.38}$$

$$\left(a_n - a_m^*\right)\delta_{mn} = 0, \quad a_n = a_n^* \tag{2.39}$$

2.4 Theorem, Hermitian Operator

For Hermitian operators when an operator A' associated with a physical observable value A acts on the state function.

$$A'\Psi = A\Psi \tag{2.40}$$

where A corresponds to a measurable physical magnitude and must be real numbers as A' is hermetic.

$$\int \psi_m^* \hat{A}\psi_n d\tau = \int \psi_n (\hat{A}\psi_m)^* d\tau \tag{2.41}$$

$$\left\langle \psi_m \left| \hat{A} \right| \psi_n \right\rangle = \left\langle \psi_n \left| \hat{A} \right| \psi_m \right\rangle^* \tag{2.42}$$

$$\int \psi_m^* \hat{A}\psi_n d\tau = \int \psi_m^* A_n \psi_n d\tau = A_n \int \psi_m^* \psi_n d\tau = A_n \int |\psi_{mn}|^2 d\tau = A_n \delta_{mn} \tag{2.43}$$

$$\int \psi_n (\hat{A}\psi_m)^* d\tau = \int \psi_n (A_m \psi_m)^* d\tau = \int \psi_n A_m^* \psi_m^* d\tau = A_m^* \int |\psi_{mn}|^2 d\tau = A_m^* \delta_{mn} \tag{2.44}$$

$$(A_n - A_m^*)\delta_{mn} = 0 \tag{2.45}$$

2.5 Theorem Orthogonality

Two eigenfunctions associated with a Hermitian operator A' with different eigenvalues are orthogonal.

Let $\hat{A}_m \psi_m = a_m \psi_m$ and $\hat{A}\psi_n = a_n \psi_n$.

If the operator A' is Hermitian:

$$\left\langle \psi_m \left| \hat{A} \right| \psi_n \right\rangle = \left\langle \psi_n \left| \hat{A} \right| \psi_m \right\rangle^* \tag{2.46}$$

$$\langle \psi_m \left| a_n \right| \psi_n \rangle = \langle \psi_n \left| a_m \right| \psi_m \rangle^* \tag{2.47}$$

$$a_n \langle \psi_m \psi_n \rangle = a_m^* \langle \psi_n \psi_m \rangle^* = a_m^* \langle \psi_m \psi_n \rangle \tag{2.48}$$

From Theorem 1, the eigenvalues of a Hermitian operator are real, $(a_m)^* = (a_m)$. As follows:

$$a_n \neq a_m \qquad \forall_{n \neq m} \quad a_n - a_m = 0 \tag{2.49}$$

$$(a_n - a_m) \langle \psi_m | \psi_n \rangle = 0 \tag{2.50}$$

$$\forall_{n \neq m} \quad \rightarrow \quad \langle \psi_m | \psi_n \rangle = 0 \tag{2.51}$$

In general,

$$\langle \psi_m \psi_n \rangle = \delta_{mn} \quad Condicion\ de\ ortonormalidad \tag{2.52}$$

2.6 Degenerate States

Let A' be an operator. The states ψ_n and ψ_m are degenerated if they have the same eigenvalue, i.e.,

$$\hat{A}\psi_n = a_n \psi_n \quad \hat{A}\psi_m = a_m \psi_m \tag{2.53}$$

with $(a_n) = (a_m)$

If A' is Hermitian, we can obtain orthogonal functions.

Let,

$$\psi_n' = \psi_n , \quad \psi_m' = \psi_n + c\psi_m \tag{2.54}$$

Orthogonality condition

$$\int \psi_n'^* \psi_m' d\tau = 1 \tag{2.55}$$

$$\int \psi_n^* (\psi_n + c\psi_m) d\tau = \int \psi_n^* \psi_n d\tau + c \int \psi_n^* \psi_m d\tau = 0 \tag{2.56}$$

Gram-Schmidt orthogonalization

$$c = -\frac{\int \psi_n^* \psi_n d\tau}{\int \psi_n^* \psi_m d\tau} \tag{2.57}$$

2.7 Linear Combination of Eigenfunctions

Let φ be a function such that

$$\varphi = \sum a_n \psi_n \tag{2.58}$$

$$\psi_m^* \varphi = \sum a_n \psi_m^* \psi_n \tag{2.59}$$

2.8 Theorem of Real Value

For Hermitian operators, when an operator A', associated with a physical observable A, acts on the state function:

$$A'\Psi = A\Psi \tag{2.60}$$

A corresponds to a measurable physical quantity, and the eigenvalues must be real, as A' is Hermitian.

$$\int \psi_m^* \hat{A} \psi_n d\tau = \int \psi_n (\hat{A} \psi_m)^* d\tau \tag{2.61}$$

$$\left\langle \psi_m \left| \hat{A} \right| \psi_n \right\rangle = \left\langle \psi_n \left| \hat{A} \right| \psi_m \right\rangle^* \tag{2.62}$$

$$\int \psi_m^* \hat{A} \psi_n d\tau = \int \psi_m^* A_n \psi_n d\tau = A_n \int \psi_m^* \psi_n d\tau = A_n \int |\psi_{mn}|^2 d\tau = A_n \delta_{mn} \tag{2.63}$$

$$\int \psi_n (\hat{A} \psi_m)^* d\tau = \int \psi_n (A_m \psi_m)^* d\tau = \int \psi_n A_m^* \psi_m^* d\tau = A_m^* \int |\psi_{mn}|^2 d\tau = A_m^* \delta_{mn} \tag{2.64}$$

$$(A_n - A_m^*) \delta_{mn} = 0 \tag{2.65}$$

2.9 Algebra of Operators

An operator is a rule that transforms a given function into another. One commonly used operator is the derivative operator $\hat{D}$. The circumflex accent will be used to identify an operator [4].

Let $f(x)$ be a function, and let $\hat{O}$ be the operator that transforms the function $f(x)$ into the function $g(x)$, written as

$$\hat{A}f(x) = g(x) \tag{2.66}$$

As examples of operators, we have Log, $\frac{d}{dx}$, $\frac{d^2}{dx^2}$, $\frac{d^n}{dx^n}$, e. We will use the notation for the derivative operator of order $\hat{D}^n$. An important property of operators is their linearity, which is a primary requirement in quantum mechanics. The linearity of an operator is defined as

2.9.1 *Linear Operators*

An operator $\hat{A}$ **is linear if**:

$$\hat{A}\left[f(x) + g(x)\right] = \hat{A}f(x) + \hat{A}g(x) \tag{2.67}$$

$$\hat{A}\left[cf(x)\right] = c\hat{A}f(x) \tag{2.68}$$

where $f(x)$ and $g(x)$ are functions and c is constant. We define the sum and difference of two operators $\hat{A}$ y $\hat{B}$ as follows:

$$\left(\hat{A} \pm \hat{B}\right) f(x) = \hat{A}f(x) \pm \hat{B}g(x) \tag{2.69}$$

The sum of linear operators has the following properties:

1. $\hat{A} + \hat{B} = \hat{B} + \hat{A}$
2. $\left(\hat{A} + \hat{B}\right) + \hat{C} = \hat{A} + \left(\hat{B} + \hat{C}\right)$
3. $\hat{A} + \hat{O} = \hat{A}, \quad \forall \hat{A}$
4. $\exists \left(-\hat{A}\right), \quad \left(-\hat{A}\right) + \hat{A} = \hat{O}$

The product of two operators $\hat{A}$ and $\hat{B}$ is defined as

$$\hat{A}\hat{B}f(x) = \hat{A}\left[\hat{B}f(x)\right] \tag{2.70}$$

The above equation indicates the order of action of the operators. In this case, the operator directly to the left of the function $f(x)$ acts first, and so on—the order of action is from right to left. The product of linear operators has the following properties:

1. $\left(\hat{A}\hat{B}\right)\hat{C} = \hat{A}\left(\hat{B}\hat{C}\right)$
2. $\hat{A}\hat{E} = \hat{E}\hat{A} = \hat{A}$
3. $\left(\hat{A} + \hat{B}\right)\hat{C} = \hat{A}\hat{C} + \hat{B}\hat{C}$
4. $\hat{A}\left(\hat{B} + \hat{C}\right) = \hat{A}\hat{B} + \hat{A}\hat{C}$

Two operators $\hat{A}$ and $\hat{B}$ are said to be equal if

$$\hat{A}f(x) = \hat{B}f(x), \quad \forall f(x) \tag{2.71}$$

2.9.2 Commutator

The commutator **Commutator** $\left[\hat{A}, \hat{B}\right]$ of two operators $\hat{A}$ and $\hat{B}$ is defined as

$$\left[\hat{A}, \hat{B}\right] = \hat{A}\hat{B} - \hat{B}\hat{A} \tag{2.72}$$

If $\hat{A}\hat{B} = \hat{B}\hat{A}$, then $\left[\hat{A}, \hat{B}\right] = 0$, and it is said that $\hat{A}$ and $\hat{B}$ are operators that commute.

Is the operator $\hat{A} = cos$? Let's check using the linearity condition of operators.

$$\hat{A}\left[c_1 f(x) + c_2 g(x)\right] = c_1 \hat{A} f(x) + c_2 \hat{A} g(x) \tag{2.73}$$

$$Cos\left[c_1 f(x) + c_2 g(x)\right] \neq c_1 Cos\left[f(x)\right] + c_2 Cos\left[g(x)\right] \tag{2.74}$$

We can write a linear differential equation using linear operators, recalling the definition of a linear differential equation, as

$$P_n(x)y^{(n)} + P_{n-1}(x)y^{(n-1)} + \ldots + P_1(x)y^{(1)} + P_0(x)y^{(0)} = Q(x) \tag{2.75}$$

With $P_n(x)$ and $Q(x)$ being functions that depend exclusively on x, the previous equation is a linear differential equation of order n, where the powers of y and its derivatives are one. Using the derivative operator $\hat{D}$, we can write the linear differential equation as

$$\left[P_n(x)\hat{D}^n + P_{n-1}(x)\hat{D}^{n-1} + \ldots + P_1(x)\hat{D}^n + P_0(x)\right]y(x) = Q(x) \tag{2.76}$$

$$\hat{L}y(x) = Q(x) \tag{2.77}$$

where

$$\hat{L}y(x) = Q(x) \tag{2.78}$$

With

$$\hat{L} = P_n(x)\hat{D}^n + P_{n-1}(x)\hat{D}^{n-1} + \ldots + P_1(x)\hat{D}^n + P_0(x) \tag{2.79}$$

Show that $\hat{L}$ is a linear operator.

2.9.3 *Eigenfunctions and Eigenvalues*

When an operator $\hat{A}$ acts on a function $f(x)$ and produces the same function multiplied by a constant λ, such that

$$\hat{A}f(x) = \lambda f(x) \tag{2.80}$$

It is said that $f(x)$ is an eigenfunction of the operator $\hat{A}$ with eigenvalue λ.

Exercise
If $f(x)$ is an eigenfunction of the linear operator $\hat{A}$ and c is a constant, show that $cf(x)$ is also an eigenfunction of the same operator with the same eigenvalue of $f(x)$.

Let's see,

$$\hat{A}\,[cf(x)] = c\hat{A}f(x) = c\lambda f(x) = \lambda\,[cf(x)] \tag{2.81}$$

Exercise

1. Let the operator $\hat{D} = \frac{d}{dx}$, obtain the functions and eigenvalues associated with it.
2. Let the operator $\hat{D}^2 = \frac{d^2}{dx^2}$, obtain the functions and eigenvalues associated with it.

For the previous exercise, consider that the eigenfunctions are finite when $x \to \pm\infty$.

Solution of the Exercise
Such as

$$\hat{A}f(x) = \lambda f(x), \quad \to \quad \hat{D}f(x) = \lambda f(x) \tag{2.82}$$

$$\frac{d}{dx}f(x) = \lambda f(x), \quad \frac{df(x)}{f(x)} = \lambda dx, \quad \ln f(x) = c + \lambda x \tag{2.83}$$

Applying the inverse function, we have

$$f(x) = Ce^{\lambda x} \tag{2.84}$$

Since λ is a complex number, then $\lambda = a + ib$.

$$f(x) = Ce^{(a+ib)x} = Ce^{ax}e^{ibx} \tag{2.85}$$

Imposing the boundary conditions $x \to \pm\infty$, we have

$$\forall a > 0, \quad x \to \infty, \quad f(x) \to \infty \tag{2.86}$$

$$\forall a < 0, \quad x \to -\infty, \quad f(x) \to \infty \tag{2.87}$$

Therefore, the boundary conditions require that $a = 0$. Thus, the eigenvalues are $\lambda = ib$, where b can be positive or negative. In this case, the eigenfunctions are

$$f(x) = c_1 e^{-ibx} + c_1 e^{+ibx} \tag{2.88}$$

2.10 Commutator

2.10.1 Definition

The commutator has been defined as

$$\left[\hat{A}, \hat{B}\right] = \hat{A}\hat{B} - \hat{B}\hat{A} \tag{2.89}$$

2.10.2 Properties of Commutators

It is defined as some properties of commutators, for two linear operators $\hat{A}$ and $\hat{B}$, and a constant k.

$$\left[\hat{A}, \hat{B}\right] = \left(\hat{A}\hat{B} - \hat{B}\hat{A}\right) = -\left(\hat{B}\hat{A} - \hat{A}\hat{B}\right) = -\left[\hat{B}, \hat{A}\right] \tag{2.90}$$

$$\left[\hat{A}, \hat{A}^n\right] = \hat{A}\hat{A}^n - \hat{A}^n\hat{A} = \hat{A}^{n+1} - \hat{A}^{n+1} = 0, \quad \forall n = 1, 2, 3, \ldots \tag{2.91}$$

$$\left[k\hat{A}, \hat{B}\right] = \left(k\hat{A}\hat{B} - \hat{B}k\hat{A}\right) = \left(\hat{A}k\hat{B} - k\hat{B}\hat{A}\right) = \left[\hat{A}, k\hat{B}\right] = k\left(\hat{A}\hat{B} - \hat{B}\hat{A}\right) = k\left[\hat{A}, \hat{B}\right] \tag{2.92}$$

$$\begin{aligned} &\left[\hat{A}, \hat{B} + \hat{C}\right] = \left(\hat{A}\left(\hat{B} + \hat{C}\right) - \left(\hat{B} + \hat{C}\right)\hat{A}\right) \\ &= \left(\hat{A}\hat{B} + \hat{A}\hat{C} - \hat{B}\hat{A} - \hat{C}\hat{A}\right) = \left(\hat{A}\hat{B} - \hat{B}\hat{A}\right) + \left(\hat{A}\hat{C} - \hat{C}\hat{A}\right) \\ &\left[\hat{A}, \hat{B} + \hat{C}\right] = \left[\hat{A}, \hat{B}\right] + \left[\hat{A}, \hat{C}\right] \end{aligned} \tag{2.93}$$

$$\begin{aligned}&\left[\hat{A}+\hat{B},\hat{C}\right]=\left(\left(\hat{A}+\hat{B}\right)\hat{C}-\hat{C}\left(\hat{A}+\hat{B}\right)\right)=\left(\hat{A}\hat{C}+\hat{B}\hat{C}-\hat{C}\hat{A}-\hat{C}\hat{B}\right)\\&=\left(\hat{A}\hat{C}-\hat{C}\hat{A}\right)+\left(\hat{B}\hat{C}-\hat{C}\hat{B}\right)\\&\left[\hat{A}+\hat{B},\hat{C}\right]=\left[\hat{A},\hat{C}\right]+\left[\hat{B},\hat{C}\right]\end{aligned}\tag{2.94}$$

$$\begin{aligned}&\left[\hat{A},\hat{B}\hat{C}\right]=\left(\hat{A}\hat{B}\hat{C}-\hat{B}\hat{C}\hat{A}\right)=\left(\hat{A}\hat{B}\hat{C}-\hat{B}\hat{A}\hat{C}-\hat{B}\hat{C}\hat{A}+\hat{B}\hat{A}\hat{C}\right)\\&=\left(\hat{A}\hat{B}-\hat{B}\hat{A}\right)\hat{C}+\hat{B}\left(\hat{A}\hat{C}-\hat{C}\hat{A}\right)\\&\left[\hat{A},\hat{B}\hat{C}\right]=\left[\hat{A},\hat{B}\right]\hat{C}+\hat{B}\left[\hat{A},\hat{C}\right]\end{aligned}\tag{2.95}$$

$$\begin{aligned}&\left[\hat{A}\hat{B},\hat{C}\right]=\left(\hat{A}\hat{B}\hat{C}-\hat{C}\hat{A}\hat{B}\right)=\left(\hat{A}\hat{B}\hat{C}-\hat{A}\hat{C}\hat{B}+\hat{A}\hat{C}\hat{B}-\hat{C}\hat{A}\hat{B}\right)\\&=\left(\hat{A}\hat{C}-\hat{C}\hat{A}\right)\hat{B}+\hat{A}\left(\hat{B}\hat{C}-\hat{C}\hat{B}\right)\\&\left[\hat{A}\hat{B},\hat{C}\right]=\left[\hat{A},\hat{C}\right]\hat{B}+\hat{A}\left[\hat{B},\hat{C}\right]\end{aligned}\tag{2.96}$$

Let's see what the result of the commutator $\left[\hat{x},\hat{p}_x\right]$ will be first, we must demonstrate that

$$\begin{aligned}&\left[\tfrac{\partial}{\partial x},\hat{x}\right]f=\tfrac{\partial}{\partial x}\hat{x}f-\hat{x}\tfrac{\partial}{\partial x}f=\tfrac{\partial}{\partial x}xf-x\tfrac{\partial f}{\partial x}=f+x\tfrac{\partial f}{\partial x}-x\tfrac{\partial f}{\partial x}=f\\&\left[\tfrac{\partial}{\partial x},\hat{x}\right]=\hat{1}\end{aligned}\tag{2.97}$$

Therefore:

$$\left[\hat{x},\hat{p}_x\right]=\left[\hat{x},-i\hbar\frac{\partial}{\partial x}\right]=-i\hbar\left[\hat{x},\frac{\partial}{\partial x}\right]=i\hbar\left[\frac{\partial}{\partial x},\hat{x}\right]=i\hbar\tag{2.98}$$

$$\left[\hat{x},\hat{p}_x^2\right]=\left[\hat{x},\hat{p}_x\right]\hat{p}_x+\hat{p}_x\left[\hat{x},\hat{p}_x\right]=i\hbar\hat{p}_x+\hat{p}_xi\hbar=2i\hbar\hat{p}_x=2i\hbar\left(-i\hbar\frac{\partial}{\partial x}\right)=2\hbar^2\frac{\partial}{\partial x}\tag{2.99}$$

$$\left[\hat{x},\hat{H}\right]=\left[\hat{x},\hat{T}+\hat{V}\right]=\left[\hat{x},\hat{T}\right]+\left[\hat{x},\hat{V}\right]=\left[\hat{x},\hat{T}\right]\tag{2.100}$$

Let's see

$$\left[\hat{x},\hat{T}\right]=\left[\hat{x},\frac{1}{2m}\left(\hat{p}_x^2+\hat{p}_y^2+\hat{p}_z^2\right)\right]=\frac{1}{2m}\left(\left[\hat{x},\hat{p}_x^2\right]+\left[\hat{x},\hat{p}_y^2\right]+\left[\hat{x},\hat{p}_z^2\right]\right)=\frac{1}{2m}\left[\hat{x},\hat{p}_x^2\right]\tag{2.101}$$

$$\left[\hat{x},\hat{T}\right]=\frac{1}{2m}2\hbar^2\frac{\partial}{\partial x}=\frac{\hbar^2}{m}\frac{\partial}{\partial x}\tag{2.102}$$

$$\left[\hat{x},\hat{H}\right]=\frac{i\hbar}{m}\left(-i\hbar\right)\frac{\partial}{\partial x}=\frac{i\hbar}{m}\hat{p}_x\tag{2.103}$$

If the measurement of a physical observable A shows a dispersion of values A_i, and $\langle A\rangle$ is the mean of these values, then the deviation of each measurement A_i from the mean is $A_i-\langle A\rangle$. The average of these deviations is zero, as we would obtain both positive and negative deviations. Therefore, to ensure all deviations are positive, we square them and call this measure the variance of A, represented in statistics by σ_A^2. In quantum mechanics, it is defined by $(\Delta A)^2$. Then:

$$(\Delta A)^2 = \langle (A_i - \langle A \rangle)^2 \rangle = \int \psi^* \left(\hat{A} - \langle A \rangle \right)^2 \psi d\tau \tag{2.104}$$

$$(\Delta A)^2 = \int \psi^* \left(\hat{A} \right)^2 \psi d\tau + \int \psi^* \left(\langle A \rangle \right)^2 \psi d\tau - 2 \int \psi^* \left(\hat{A} \langle A \rangle \right) \psi d\tau \tag{2.105}$$

$$(\Delta A)^2 = \langle A^2 \rangle - \langle A \rangle^2 \tag{2.106}$$

The standard deviation σ is defined as the square root of the variance $(\Delta A)^2$. The standard deviation of the physical observable A represents the measurement uncertainty of that observable.

$$\sigma = \sqrt{\langle A_i^2 \rangle - \langle A \rangle^2} \tag{2.107}$$

Now, let us consider two physical observables, A and B. The product of the standard deviations of the two observables satisfies the following condition:

$$\Delta A \Delta B \geq \frac{1}{2} \left| \int \psi^* \left[\hat{A}, \hat{B} \right] \psi d\tau \right| \tag{2.108}$$

$$\Delta x \Delta p_x \geq \frac{1}{2} \left| \int \psi^* \left[\hat{A}, \hat{B} \right] \psi d\tau \right| = \frac{1}{2} \left| \int \psi^* \left[\hat{x}, \hat{p}_x \right] \psi d\tau \right| = \frac{1}{2} \left| \int \psi^* i\hbar \psi d\tau \right| = \frac{1}{2} |i| \hbar \tag{2.109}$$

Since complex numbers do not form an ordered set in the same way as real numbers, generalizing this concept is not straightforward. It requires the following identity, which provides an alternative and equivalent definition of the absolute value:

$$Z = a + ib \tag{2.110}$$

With a and b being real numbers, the absolute value or modulus of z is formally defined by

$$|Z| = Z \times Z^* = \sqrt{a^2 + b^2} \tag{2.111}$$

The geometric interpretation of the absolute value for real and complex numbers can be related to the Pythagorean theorem, where the absolute value of a complex number corresponds to the distance from that number to the origin in the complex plane. The absolute value of the difference between two complex numbers is equal to the distance between them. Therefore:

$$|i| = \sqrt{a^2 + b^2} = \sqrt{0^2 + 1^2} = 1 \tag{2.112}$$

Therefore

$$\Delta x \Delta p_x \geq \frac{1}{2}\hbar \tag{2.113}$$

The above inequality is known as the Heisenberg uncertainty relation.

Let us see an application of the above relations:
Consider a particle in a one-dimensional potential where the wave equation is given as

$$\psi_n(x) = \begin{cases} Asen\frac{n\pi}{a}x & a \geq x \geq 0 \\ 0 & a < x < 0 \end{cases} \tag{2.114}$$

where $A_n = \sqrt{\frac{2}{a}}$.

Let us calculate the mean values (or expected values) of $\langle x \rangle$ and $\langle x \rangle$ for the state $n = 1$.

$$\langle x \rangle = \int \psi^* \hat{x} \psi d\tau = \int_0^a \psi^* \hat{x} \psi dx = \int_0^a x \left(\sqrt{\frac{2}{a}} sen \left(\frac{n\pi}{a} x \right) \right)^2 dx \tag{2.115}$$

$$\langle x \rangle = \frac{2}{a} \int_0^a xsen \left(\frac{\pi}{a} x \right)^2 dx = \frac{a}{2} \tag{2.116}$$

where the following integral has been used:

$$\int xsen^2(cx)\,dx = \frac{x^2}{4} - \frac{x}{4c} sen(2cx) - \frac{1}{8c^2}\cos(2cx) \tag{2.117}$$

Then,

$$\int_0^a xsen^2 \left(\frac{\pi}{a} x \right) dx = \frac{x^2}{4} - \frac{x}{4c} sen(2cx) - \frac{1}{8c^2}\cos(2cx) \Big|_0^a \tag{2.118}$$

$$\int_0^a xsen^2 \left(\frac{\pi}{a} x \right) dx = \frac{a^2}{4} - 0 - \frac{a^2}{4\pi} sen(2\pi) + \frac{a^2}{4\pi} sen(0) - \frac{a^2}{8\pi^2}\cos(2\pi) + \frac{a^2}{8\pi^2}\cos(0) \tag{2.119}$$

$$\int_0^a xsen^2 \left(\frac{\pi}{a} x \right) dx = \frac{a^2}{4} \tag{2.120}$$

$$\langle x \rangle = \frac{a}{2} \tag{2.121}$$

Let's see how much is the standard deviation for the linear momentum in the x-direction

$$\langle p_x\rangle = \int \psi^* \hat{p}_x \psi d\tau = \int_0^a \psi^* \left(-i\hbar\frac{\partial}{\partial x}\right)\psi dx = \int_0^a \sqrt{\frac{2}{a}} sen\left(\frac{\pi}{a}x\right)\left(-i\hbar\frac{\partial}{\partial x}\right)\left(\sqrt{\frac{2}{a}} sen\left(\frac{\pi}{a}x\right)\right)dx \tag{2.122}$$

$$\langle p_x\rangle = -\frac{2i\hbar}{a}\int_0^a \frac{\pi}{a} sen\left(\frac{\pi}{a}x\right)\cos\left(\frac{\pi}{a}x\right)dx = 0 \tag{2.123}$$

Task demonstrates that

$$\langle x^2\rangle = a^2\left(\frac{1}{3} - \frac{1}{2\pi^2}\right) \tag{2.124}$$

Then

$$(\Delta x)^2 = \langle x^2\rangle - \langle x\rangle^2 = a^2\left(\frac{1}{3} - \frac{1}{2\pi^2}\right) - \frac{a^2}{4} = a^2\left(\frac{\pi^2-6}{12\pi^2}\right) \tag{2.125}$$

$$\Delta x = \frac{a}{\pi}\left(\frac{\pi^2-6}{12}\right)^{\frac{1}{2}} \tag{2.126}$$

Now for Δp_x, show that

$$\langle p_x^2\rangle = \frac{h^2}{4a^2} \tag{2.127}$$

Then

$$(\Delta p_x)^2 = \langle p_x^2\rangle - \langle p_x\rangle^2 = \frac{h^2}{4a^2} \tag{2.128}$$

$$\Delta p_x = \frac{h}{2a} \tag{2.129}$$

$$\Delta x \Delta p_x = \frac{a}{\pi}\left(\frac{\pi^2-6}{12}\right)^{\frac{1}{2}}\frac{h}{2a} = \frac{h}{2\pi}\left(\frac{\pi^2-6}{12}\right)^{\frac{1}{2}} = 0.568\hbar > \frac{1}{2}\hbar \tag{2.130}$$

Considering that the energy operator must be re-evaluated, i.e., it is not $i\hbar\frac{\partial}{\partial t}$, but the Hamiltonian operator, and that time t is a parameter (i.e., it is not a physical observable), show that

$$\Delta E \Delta t \geq \frac{1}{2}\hbar \tag{2.131}$$

2.11 Proposed Problems

2.1 Demonstrate that two eigenfunctions associated with a Hermitian operator $\hat{A}$, corresponding to different eigenvalues, are orthogonal.

2.2 Consider a particle in a one-dimensional potential with infinite walls. Obtain the eigenfunctions and eigenvalues.

2.3 Consider a particle in a one-dimensional potential with finite height V_w and width L. Obtain the eigenfunctions and eigenvalues.

References

1. Dirac PAM (1981) The principles of quantum mechanics, 4th ed. Oxford University Press
2. Zettili N (2009) Quantum mechanics: concepts and applications. Wiley
3. Sakurai JJ (1994) Modern quantum mechanics (Revised ed). Addison-Wesley
4. Louisell, William H (1973) Quantum statistical properties of radiation. Wiley

Chapter 3
Mathematical Fundamentals of Perturbation Theory

Abstract Perturbation theory is a powerful method used in quantum mechanics to find an approximate solution to a problem that cannot be solved exactly. It is particularly useful when dealing with a Hamiltonian that can be separated into a solvable part and a small perturbing part. This theory is applied when the perturbations are smaller than the unperturbed system.

3.1 Time-Independent Perturbation Theory

Let's consider the Hamiltonian [1]:

$$H = H_0 + H' \tag{3.1}$$

where the eigenstates H_0 and eigenvalues E_0 are known, and H' is much smaller than H_0. Additionally, H_0 and H' are time independent. In this case, we define the Hamiltonians as follows:

$H_0 =$ Unperturbed Hamiltonian.

$H' =$ Perturbation Hamiltonian.

The problem to solve is finding the modifications to the energy levels of the system H in a stationary state due to the addition of a perturbation.

Let $|\Psi_p^i\rangle$ be the known wave functions of the system H_0 that form an orthonormal basis satisfying

$$\langle\Psi_p^i|\Psi_{p'}^{i'}\rangle = \delta_{pp'}\delta_{ii'} \qquad \sum_p\sum_i |\Psi_p^i\rangle\langle\Psi_p^i| = 1 \tag{3.2}$$

where p and i are integers that label the state and the degeneracy; moreover, each wave function is associated with an energy E_p^0, i.e.:

$$|\Psi_p^i\rangle \quad \rightarrow \quad E_p^0 \tag{3.3}$$

C. Vargas Hernandez, *Introduction to Raman Spectroscopy and Its Applications*,
https://doi.org/10.1007/978-3-031-77551-2_3

Since the unperturbed system H_0 is known, we have the eigenvalue equation:

$$H_0|\Psi_p^i\rangle = E_p^0|\Psi_p^i\rangle \tag{3.4}$$

It introduces a strategy to measure the size of the perturbation relative to the unperturbed system, and we will call this indicator the order parameter λ. For our case, where we consider a small perturbation applied to the system such that $\lambda << 1$, then the Hamiltonian of the system H can be written as

$$H = H_0 - \lambda H' \tag{3.5}$$

The effect of introducing a perturbation from a phenomenological point of view is to lift the degeneracy of the system regarding the energy of the corresponding eigenvalue of H_0. This means that the system to solve is

$$H|\Psi\rangle = E|\Psi\rangle \quad \Rightarrow \quad H(\lambda)|\Psi(\lambda)\rangle = E(\lambda)|\Psi(\lambda)\rangle \tag{3.6}$$

Assuming that $|\Psi\rangle$ and E can be expanded in power series of λ as follows:

$$E(\lambda) = \varepsilon_0 + \lambda\varepsilon_1 + \lambda^2\varepsilon_2 + ...\lambda^q\varepsilon_q + ... \tag{3.7}$$

$$\Psi(\lambda)\rangle = |0\rangle + \lambda|1\rangle + \lambda^2|2\rangle + ...\lambda^q|q\rangle + ... \tag{3.8}$$

Substituting 3.7 into 3.6, we have

$$(H_0 + \lambda H')\ \ [|0\rangle + \lambda|1\rangle + \lambda^2|2\rangle + ...\lambda^q|q\rangle + ...] = \tag{3.9}$$

$$(\varepsilon_0 + \lambda\varepsilon_1 + \lambda^2\varepsilon_2 + ...\lambda^q\varepsilon_q + ...)\ \ [|0\rangle + \lambda|1\rangle + \lambda^2|2\rangle + ...\lambda^q|q\rangle + ...] \tag{3.10}$$

$$(H_0 + \lambda H')\ \ [\sum_{q=0}^{\infty}\lambda^q|q\rangle] = [\sum_{l=0}^{\infty}\lambda^l\varepsilon_l][\sum_{q=0}^{\infty}\lambda^q|q\rangle] \tag{3.11}$$

Separating terms in the respective powers of λ:

$$H_0|0\rangle + \lambda H_0|1\rangle + \lambda^2 H_0|2\rangle + ...\lambda^q H_0|q\rangle + \lambda H'|0\rangle + \lambda^2 H'|1\rangle + ...\lambda^{q+1} H'|q\rangle = \tag{3.12}$$

$$[\varepsilon_0|0\rangle + \lambda\varepsilon_0|1\rangle + \lambda^2\varepsilon_0|2\rangle + ...\lambda^q\varepsilon_0|q\rangle + \lambda\varepsilon_1|0\rangle + \lambda^2\varepsilon_1|1\rangle + ...\lambda^{q+1}\varepsilon_q|q\rangle + \lambda^2\varepsilon_2|0\rangle + \lambda^3\varepsilon_2|1\rangle + ...\lambda^{q+2}\varepsilon_q|q\rangle + ...] \tag{3.13}$$

Equalizing the respective powers of λ, we obtain the solution for zero order, first order, and second order as follows:

3.1.1 *Zero Order*

$$H_0|0\rangle = \varepsilon_0|0\rangle \tag{3.14}$$

3.1.2 *First Order*

$$\lambda H_0|1\rangle + \lambda H'|0\rangle = \lambda\varepsilon_0|1\rangle + \lambda\varepsilon_1|0\rangle \tag{3.15}$$

$$\boxed{(H_0 - \varepsilon_0)|1\rangle + (H' - \varepsilon_1)|0\rangle = 0} \tag{3.16}$$

3.1.3 *Second Order*

$$\lambda^2 H_0|2\rangle + \lambda^2 H'|1\rangle = \lambda^2\varepsilon_0|2\rangle + \lambda^2\varepsilon_1|1\rangle + \lambda^2\varepsilon_2|0\rangle \tag{3.17}$$

$$\boxed{(H_0 - \varepsilon_0)|2\rangle + (H' - \varepsilon_1)|1\rangle - \varepsilon_2|0\rangle = 0} \tag{3.18}$$

3.1.4 *Terms of Order Q*

$$\boxed{(H_0 - \varepsilon_0)|q\rangle + (H' - \varepsilon_1)|q-1\rangle - \varepsilon_2|q-2\rangle - ...\varepsilon_q|0\rangle = 0} \tag{3.19}$$

The set $|q\rangle$ is considered orthonormalized, then

$$\langle\Psi(\lambda)|\Psi(\lambda)\rangle = [\langle 0| + \lambda\langle 1| + ...O(\lambda^p)][|0\rangle + \lambda|1\rangle + O'(\lambda^p)] \tag{3.20}$$

With $p > 1$ $O \equiv$ Means the rest of the terms
Since $\langle 0|0\rangle = 1$ Orden Zero order
$\langle 0|1\rangle = \langle 1|0\rangle = 0$ For first order

$$\langle\Psi(\lambda)|\Psi(\lambda)\rangle = \langle 0|0\rangle + \lambda[\langle 1|0\rangle + \langle 0|1\rangle] + O^2(\lambda^p) \tag{3.21}$$

Since $|0\rangle$ is an eigenvector of H_0

$$\langle 0|0\rangle = 1 \quad \Rightarrow \quad \langle 0|1\rangle = \langle 1|0\rangle = 0 \tag{3.22}$$

To second order λ^2

$$1 = \langle 0|2\rangle + \langle 2|0\rangle + \langle 1|1\rangle, \quad \langle 0|2\rangle = \langle 2|0\rangle \tag{3.23}$$

$$\boxed{\langle 0|2\rangle = -\frac{1}{2}\langle 1|1\rangle} \tag{3.24}$$

3.2 Perturbation Theory for Non-Degenerate Energy Levels

Consider a non-degenerate energy level E_p^0 of the unperturbed Hamiltonian H_0 and let the associated wave function be

$$|\Psi_p^0\rangle \equiv |\Psi_p\rangle \quad \wedge \quad E_p^0 \equiv E_p \tag{3.25}$$

When

$\lambda \rightarrow 0 \Rightarrow E_p^0 \equiv \varepsilon$ according to Eq. 3.1.1

$$\Rightarrow \quad |0\rangle = |\Psi_p\rangle = |p\rangle \tag{3.26}$$

3.2.1 First-Order Energy Correction

As

$$(H_0 - \varepsilon_0)|1\rangle + (H' - \varepsilon_1)|0\rangle = 0 \tag{3.27}$$

Projecting this equation onto the vector $|0\rangle$ and using the projector $|0\rangle\langle 0| = 1$

$$\langle 0|(H_0 - \varepsilon_0)|0\rangle\langle 0|1\rangle + \langle 0|(H' - \varepsilon_1)|0\rangle\langle 0|0\rangle = 0 \tag{3.28}$$

$$\Rightarrow \quad \varepsilon_1 = \langle 0|H'|0\rangle \tag{3.29}$$

the energy correction for the level p is E_p

$$\boxed{\varepsilon_1 = \langle 0|H'|0\rangle} \tag{3.30}$$

Then

$$\boxed{E_p = E_p^0 + \langle 0|H'|0\rangle} \tag{3.31}$$

$E_1^0 \equiv$ Energy level of the unperturbed system with $|p\rangle$ the eigenstates of H_0.

3.2.2 First-Order Correction of Eigenstates

Let $|0\rangle \equiv |p\rangle$ be the unperturbed state with energy E_p^0. For the state $|p\rangle$ we want to determine the first-order correction, then

$$(H_0 - E_p^0)|1\rangle + (H' - \varepsilon_1)|p^0\rangle = 0 \tag{3.32}$$

Projecting onto the vector $|n^0\rangle$

$$\langle n^0|(H_0 - E_p^0)|1\rangle + \langle n^0|(H' - \varepsilon_1)|p^0\rangle = 0 \qquad p \neq n \tag{3.33}$$

$$\langle n^0|(H_0 - E_p^0)|n^0\rangle\langle n^0|1\rangle + \langle n^0\, H'|p^0\rangle = 0 \tag{3.34}$$

$$(E_n^0 - E_p^0)\langle n^0|1\rangle + \langle n^0\, H'|p^0\rangle = 0 \tag{3.35}$$

$$\boxed{\langle n^0|1\rangle = \frac{1}{E_n^0 - E_p^0}\langle n^0\, H'|p^0\rangle} \tag{3.36}$$

It can write $|1\rangle$ as a linear combination of $|n^0\rangle$

$$\boxed{|1\rangle = \sum_n a_n |n^0\rangle} \tag{3.37}$$

$$\langle m^0|1\rangle = \sum_n a_n \langle m^0|n^0\rangle \tag{3.38}$$

$$\langle m^0|1\rangle = \sum_n a_n \delta^0_{mn} = a_m \tag{3.39}$$

$$\boxed{a_m = \langle m^0|1\rangle} \tag{3.40}$$

in 3.36

$$a_n = \frac{1}{E^0_n - E^0_p} \langle n^0\, H'|p^0\rangle \tag{3.41}$$

then in 3.37

$$\boxed{|1\rangle^p = \sum_{n \neq p} (\frac{\langle n^0\, H'|p^0\rangle}{E^0_n - E^0_p})|n^0\rangle} \tag{3.42}$$

$|1\rangle^p \equiv$ First-order correction of the unperturbed state $|p\rangle$, then in 3.7

$$\boxed{|\Psi_p\rangle^p = |p^0\rangle + |1\rangle^p} \tag{3.43}$$

3.2.3 *First-Order Wavefunction Correction in the First State*

$$|p\rangle = |0\rangle \tag{3.44}$$

$$|\Psi_0\rangle = |0^0\rangle + |1\rangle^0 \tag{3.45}$$

$$|1\rangle^0 = \sum_{n \neq 0} \frac{\langle n^0\, H'|0\rangle}{E^0_0 - E^0_n} |n^0\rangle \tag{3.46}$$

3.2.4 *Second-Order Energy Corrections*

$$(H_0 - E_n^0)|2\rangle + (H' - \varepsilon_1)|1\rangle - \varepsilon_2|n^0\rangle = 0 \tag{3.47}$$

$$\langle p^0|(H_0 - E_n^0)|2\rangle + \langle p^0|(H' - \varepsilon_1)|1\rangle - \varepsilon_2\langle p^0|n^0\rangle = 0 \tag{3.48}$$

$$\langle p^0|(H_0 - E_n^0)|p^0\rangle\langle p^0|2\rangle + \langle p^0\, H'|1\rangle - \varepsilon_1\langle p^0|1\rangle - \varepsilon_2\delta^0_{pn} = 0 \tag{3.49}$$

$$(E_p^0 - E_n^0)\langle p^0|2\rangle + \langle p^0\, H'|1\rangle = \varepsilon_2\delta_{pn} \tag{3.50}$$

$$\boxed{\varepsilon_2 = \langle n^0\, H'|1\rangle} \tag{3.51}$$

but

$$|1\rangle = \sum_{n\neq p}\left(\frac{\langle n^0\, H'|p^0\rangle}{E_p^0 - E_n^0}\right)|n^0\rangle \tag{3.52}$$

then substituting

$$\varepsilon_2 = \langle p^0\, H'[\sum_{n\neq p}\left(\frac{\langle n^0\, H'|p^0\rangle}{E_p^0 - E_n^0}\right)]|n^0\rangle \tag{3.53}$$

$$\varepsilon_2 = \sum_{n\neq p}\left(\frac{\langle n^0\, H'|p^0\rangle}{E_p^0 - E_n^0}\right)\langle p^0\, H'|n^0\rangle \tag{3.54}$$

$$\boxed{\varepsilon_2 = \sum_{n\neq p}\frac{|\langle n^0\, H'|p^0\rangle|^2}{E_p^0 - E_n^0}} \tag{3.55}$$

3.2.5 *Example The Stark Effect*

The Stark effect is the shift in the energy levels of an atom or molecule due to an external electric field. In this practical example, we consider the hydrogen atom in an external electric field in the z-direction.

Unperturbed Hamiltonian
Consider that the Hamiltonian $H_0$for a hydrogen atom without a perturbation is known, and it is defined as

$$\hat{H}_0 = -\frac{\hbar^2}{2m}\nabla^2 - \frac{q^2}{4\pi\varepsilon_0 r} \tag{3.56}$$

where the first and second terms on the right side of the equation are associated with kinetic and potential energies, respectively.

The eigenstates and eigenvalues of $\hat{H}_0$are well known:

$$\hat{H}_0\psi_{n,l,m} = E_n\psi_{n,l,m} \tag{3.57}$$

where

$$E_n = \frac{-13.6Z_{efect}^2}{n^2}eV \tag{3.58}$$

Perturbing Hamiltonian The perturbing Hamiltonian due to an external electric field E in the z-direction is

$$\hat{H}\prime = eEz \tag{3.59}$$

First-Order Energy Correction To find the first-order correction to the energy levels, we use

$$E_n^{(1)} = \left\langle \psi_{n,l,m} \right| \hat{H}\prime \left| \psi_{n,l,m} \right\rangle \tag{3.60}$$

For the hydrogen atom, the wave functions $\psi_{n,l,m}$ are orthogonal and the selection rules for electric dipole transitions require $\Delta l = \pm 1$. Therefore, only certain states will have non-zero matrix elements.

Consider the ground state ($n = 1, l = 0, m = 0$):

$$\psi_{1,0,0} = \frac{1}{\sqrt{\pi a_0^3}}e^{-\frac{r}{a_0}} \tag{3.61}$$

where a_0 is the Bohr radius.

The first-order energy correction for the ground state is

$$E_1^{(1)} = \left\langle \psi_{1,0,0} \right| eEz \left| \psi_{1,0,0} \right\rangle \tag{3.62}$$

Since $z = rcos\theta$ and the ground state wave function is spherically symmetric, the integral over $cos\theta$ will be zero. Thus:

$$E_1^{(1)} = 0 \tag{3.63}$$

Second-Order Energy Correction To find the second-order energy correction, we use

$$E_n^{(2)} = \sum_{k \neq n} \frac{|\langle \psi_k | eEz | \psi_n \rangle|^2}{E_n^{(0)} - E_k^{(0)}} \tag{3.64}$$

For the ground state, this becomes

$$E_1^{(2)} = \sum_{n \neq 1} \frac{|\langle \psi_1 | eEz | \psi_n \rangle|^2}{E_1^{(0)} - E_n^{(0)}} \tag{3.65}$$

Since the first-order correction is zero, the dominant effect of the external field on the ground state energy comes from the second-order term. In this example, perturbation theory allows us to calculate the shift in the energy levels of the hydrogen atom due to an external electric field (Stark effect). The first-order correction is zero for the ground state, so the second-order correction provides the leading contribution to the energy shift. This approach is widely applicable in quantum mechanics to study the effects of small perturbations on a system's energy levels and wave functions.

3.2.6 *Example Harmonic Oscillator*

Consider the one-dimensional harmonic oscillator with:

$$H_0 = \frac{p^2}{2m} + \frac{1}{2} m\omega^2 x^2 \qquad with \qquad E_n^0 = (n + \frac{1}{2})\hbar\omega \tag{3.66}$$

Consider the perturbation $\gamma_1 x^3 + \gamma_2 x^4$, then

$$\boxed{\hat{H} = \frac{p^2}{2m} + \frac{1}{2} m\omega^2 x^2 + \gamma_1 x^3 + \gamma_2 x^4} \tag{3.67}$$

then

$$E_p = E_p^0 + \langle p | H' | p \rangle \tag{3.68}$$

for the ground state of the harmonic oscillator

$$|0^0\rangle = (\frac{\alpha}{\pi})^{\frac{1}{4}} e^{-\alpha\frac{x^2}{2}}, \qquad \alpha = \frac{2\pi \nu m}{\hbar} \tag{3.69}$$

then

$$E_0 = E_0^0 + \langle 0^0\, H'|0^0\rangle \tag{3.70}$$

$$\varepsilon_1 = \langle 0^0\, H'|0^0\rangle = \int_{-\alpha}^{\alpha} \left(\frac{\alpha}{\pi}\right)^{\frac{1}{2}} e^{-\alpha x^2} \left(\gamma_1 x^3 + \gamma_2 x^4\right) \tag{3.71}$$

the function $e^{-\alpha x^2} x^3$ is odd, then

$$\int_{-\alpha}^{\alpha} (\frac{\alpha}{\pi})^{\frac{1}{2}} \gamma_1 e^{-\alpha x^2} x^3 \quad \to \quad 0 \tag{3.72}$$

and since $e^{-\alpha x^2} x^4$ is even

$$\varepsilon_1 = 2\gamma_2 (\frac{\alpha}{\pi})^{\frac{1}{2}} \int_0^{\alpha} e^{-\alpha x^2} x^4 \tag{3.73}$$

knowing that

$$\int_0^{\alpha} x^{2n} e^{-bx^2} dx = \frac{1 * 3 ... (2n-1)}{2^{n+1}} (\frac{\pi}{b^{2n+1}})^{\frac{1}{2}}, \qquad b > 0, \qquad n = 1, 2, 3, ... \tag{3.74}$$

then

$$\varepsilon_1 = 2\gamma_2 (\frac{\alpha}{\pi})^{\frac{1}{2}} \int_0^{\alpha} x^4 e^{-\alpha x^2} dx = 2\gamma_2 (\frac{\alpha}{\pi})^{\frac{1}{2}} [\frac{1 * 3}{2^3} (\frac{\pi}{\alpha^5})^{\frac{1}{2}}] \ , \qquad \alpha > 0 \tag{3.75}$$

let $n = 2 \wedge b = \alpha$

$$\varepsilon_1 = \frac{3\gamma_2}{4\alpha^2}, \qquad E_n^0 = (n + \frac{1}{2})\hbar\omega, \qquad \alpha = \frac{2\pi \nu m}{\hbar} = \frac{4\pi^2 m}{h^2}\hbar\omega \tag{3.76}$$

$$\boxed{E_0 = \frac{1}{2}\hbar\omega + \frac{3\gamma_2}{4\alpha^2}} \tag{3.77}$$

$\forall n = 0$ (Ground state)

3.3 Time-Dependent Perturbation Theory

When matter interacts with electromagnetic radiation, absorption and the respective emission of photons may occur, producing an energy dispersion spectrum, among which we can find transitions between atomic or molecular levels. This incident electromagnetic radiation has an oscillating time-dependent field, which produces a time-dependent energy dispersion in the interaction with matter. Therefore, it is necessary to use time-dependent perturbation theory. To solve a time-dependent quantum system, the time-dependent Schrödinger equation must be solved.

$$\hat{H}\,|N,t\rangle = -\frac{\hbar}{i}\frac{d}{dt}\,|N,t\rangle \tag{3.78}$$

Similar to time-independent perturbation theory, we now consider the perturbation as a function that depends on time, so that we can consider the total Hamiltonian of the system as the sum of the stationary Hamiltonian $\hat{H}_0$ and the perturbation $V(t)$, which corresponds to the external interaction potential.

That is:

$$\hat{H} = \hat{H}_0 + V(t) \tag{3.79}$$

Knowing that we know the solution of the unperturbed quantum system; that is, we know the eigenfunctions $|n,t\rangle$ and the respective eigenvalues of the stationary problem, so we know the solution of the equation:

$$\hat{H}_0\,|n,t\rangle = -\frac{\hbar}{i}\frac{d}{dt}\,|n,t\rangle \tag{3.80}$$

where the solution of the above equation produces eigenfunctions whose functional expression is

$$|n,t\rangle = e^{\frac{iE_n t}{\hbar}}\,|n\rangle \tag{3.81}$$

These eigenfunctions form an orthogonal basis that constitutes a good set, which can be used to find the eigenfunctions of the time-dependent Hamiltonian, that is, we can think that the eigenfunctions of $\hat{H}$ can be constructed as a linear combination of the eigenfunctions $|n, t\rangle$ of the Hamiltonian $\hat{H}_0$, that is:

$$|N, t\rangle = \sum_n C_n(t) e^{\frac{iE_n t}{\hbar}} |n\rangle \tag{3.82}$$

Substituting in Eq. (2.12) Eqs. (2.13) and (2.16), we obtain

$$\sum_n C_n(t) \hat{H}_0 |n, t\rangle + \sum_n C_n(t) e^{-\frac{iE_n}{\hbar}} V(t) |n\rangle = -\frac{\hbar}{i} \sum_n \frac{dC_n(t)}{dt} e^{-\frac{iE_n}{\hbar}} |n\rangle - \frac{\hbar}{i} \sum_n C_n(t) \frac{\partial}{\partial t} |n, t\rangle \tag{3.83}$$

By multiplying the previous equation by a bracket defined as

$$\langle p, t| = \langle p| e^{\frac{iE_p t}{\hbar}} \tag{3.84}$$

one obtains

$$\sum_n C_n(t) E_n e^{\frac{i(E_p - E_n)t}{\hbar}} \langle p \mid n\rangle + \sum_n C_n(t) e^{\frac{i(E_p - E_n)t}{\hbar}} \langle p| V(t) |n\rangle = -\frac{\hbar}{i} \sum_n \frac{dC_n(t)}{dt} e^{\frac{i(E_p - E_n)t}{\hbar}} \langle p \mid n\rangle \tag{3.85}$$

Due to the orthogonality of the wave functions of the stationary system, i.e., $\langle p \mid n\rangle = \delta_{pn}$. Applying the necessary algebra and rearranging, we obtain

$$\sum_n C_n(t) e^{\frac{i(E_p - E_n)t}{\hbar}} \langle p| V(t) |n\rangle = -\frac{\hbar}{i} \sum_n \frac{dC_n(t)}{dt} \tag{3.86}$$

Taking $\langle p| V(t) |n\rangle = V_{pn}$ we write the equation:

$$\sum_n C_n(t) e^{\frac{i(E_{pn})t}{\hbar}} V_{pn} = -\frac{\hbar}{i} \sum_n \frac{dC_n(t)}{dt} \tag{3.87}$$

In the solution of the previous equation, initial conditions must be considered. For $t < 0$, it is considered that the system is in a state $|i\rangle$, which means that of all the coefficients $C_n(t)$, only the coefficient $C_i(t)$ is different from zero and independent of $C_n(t)$. The temporal dependence of the coefficients $C_n(t)$ is associated with the evolution of the system. Since the solution of the equation is continuous at $t = 0$ and $V(t)$ is finite, then

$$C_n(t = 0) = \delta_{ni} \tag{3.88}$$

If $V(t)$ is very small compared to $\hat{H}_0$, in the first approximation the initial state $|i\rangle$ will be very little perturbed, the system will evolve slightly, and the coefficients $C_n(t)$ can be considered constant. Then:

$$C_n(t = 0) = C_n(t) = \delta_{ni} \tag{3.89}$$

and we have that

$$i\hbar \frac{dC_p(t)}{dt} = \sum \delta_{ni} e^{\frac{iE_{pn}t}{\hbar}} V_{pn} \tag{3.90}$$

which yields the expression:

$$i\hbar \frac{dC_p(t)}{dt} = e^{\frac{iE_{pn}t}{\hbar}} V_{pi} \tag{3.91}$$

The previous equation can be integrated, resulting in

$$C_p(t) = \frac{1}{i\hbar} \int_0^t e^{\frac{iE_{pi}t}{\hbar}} V_{pi}(t)dt \tag{3.92}$$

We define the transition probability between an initial state i and a final state f as

$$P_{if}(t) = \left|C_f(t)\right|^2 \tag{3.93}$$

Substituting accordingly:

$$P_{if}(t) = \frac{1}{\hbar^2} \left| \int_0^t e^{\frac{iE_{pi}t}{\hbar}} V_{pi}(t)dt \right|^2 \tag{3.94}$$

Consider that the perturbation potential is known and has the form:

$$V(t) = V_{0if} Cos(\omega t) = \frac{V_{0if}}{2i}(e^{i\omega t} - e^{-i\omega t}) \tag{3.95}$$

where it is assumed that V_{0if} is time independent and ω is the frequency of the external signal perturbing the system. This means that $P_{if}(t)$ represents the transition probability from the initial state i to the final state f induced by the external perturbation, which mathematically can represent an electromagnetic wave of frequency ω.

If we substitute Eq. 3.118 into Eq. 3.92, we obtain

$$C_p(t) = \frac{1}{i\hbar} \int_0^t e^{\frac{iE_{pn}t}{\hbar}} \langle p| \frac{V_{0if}}{2i}(e^{i\omega t} - e^{-i\omega t}) |n\rangle \, dt \tag{3.96}$$

resulting in

$$C_p(t) = \frac{V_{0if}}{i\hbar} \int_0^t \left[e^{i(\omega_{pi}+\omega)t} - e^{i(\omega_{pi}-\omega)t}\right] dt \tag{3.97}$$

Performing the integration yields

$$C_p(t) = \frac{V_{oif}}{2i\hbar} \left(\frac{1 - e^{i(\omega_{pi}+\omega)t}}{\omega_{pi} + \omega} - \frac{1 - e^{i(\omega_{pi}+\omega)t}}{\omega_{pi} - \omega} \right) \tag{3.98}$$

Then, the transition probability between the initial state $|i\rangle$ and the final state $|p\rangle$ induced by the external perturbation $V(t) = \frac{V_{oif}}{2i}(e^{i\omega t} - e^{-i\omega t})$, according to Eq. 3.93, is

$$P_{pi}(t, \omega) = \frac{\left|V_{opi}\right|^2}{4\hbar^2} \left[\frac{1 - e^{i(\omega_{pi}+\omega)t}}{\omega_{pi} + \omega} - \frac{1 - e^{i(\omega_{pi}+\omega)t}}{\omega_{pi} - \omega} \right] \tag{3.99}$$

The transition probability is maximized in the resonance region, that is, when

$$\omega_{pi} + \omega = 0 \tag{3.100}$$

$$\omega = -\omega_{pi} = \omega_i - \omega_p > 0 \tag{3.101}$$

$$\hbar\omega = -\hbar\omega_{pi} = \hbar\omega_i - \hbar\omega_p > 0 \tag{3.102}$$

$$E_i > E_p \tag{3.103}$$

Since the energy in the final state $|p\rangle$ is less than the energy in the initial state $|i\rangle$, this transition is called an emission process. In the other resonant case, we have that

$$P_{pi}(t,\omega) = \left|\frac{V_{opi}}{\hbar}\right|^2 \frac{sen^2\left(\frac{1}{2}(\omega - \omega_{pi})t\right)}{(\omega - \omega_{pi})^2} \tag{3.104}$$

This equation explains the elastic process called Rayleigh scattering, which indicates that when a photon strikes a material characterized by two quantum states; $|i\rangle$ and $|p\rangle$, the transition occurs, resulting in absorption and, after a time called the relaxation time (on the order of $10^{-9}s$), re-emission occurs. The inelastic process called Raman scattering can be modeled by a virtual state and two photons, requiring the use of the quantum description of perturbation theory. Time-dependent perturbation theory provides a powerful framework for analyzing the effects of time-varying external fields on molecular systems. By applying this theory to a diatomic molecule in an oscillating electric field, we gain insights into vibrational transitions and resonance phenomena, essential for various spectroscopic applications.

3.3.1 Example

For dynamic systems, i.e., those systems whose parameters change over time, it is said that the system evolves. Time-dependent perturbation theory is used in quantum mechanics to describe how a quantum system evolves when subject to a time-dependent perturbation. In many cases, the perturbation is an electromagnetic wave that interacts with the system and generates transitions between different quantum states. In the following example, the solution for this interaction is carried out.

Consider a two-level quantum system with states $|\psi_1\rangle$ and $|\psi_2\rangle$, with corresponding energies E_1 and E_2. Initially, the system is in state $|\psi_1\rangle$. A time-dependent perturbation $H'(t) = V_0 e^{-i\omega t}$ is applied, where V is a small perturbation compared with the system energy unperturbed and ω is the frequency of the perturbation.

The question is how to determine the probability that the system transitions from state $|\psi_1\rangle$ to state $|\psi_2\rangle$ at time t. In this case, the total Hamiltonian of the system has two terms: the unperturbed Hamiltonian and the perturbed Hamiltonian.

$$H(t) = H_0 + H'(t) \tag{3.105}$$

where H_0 is the unperturbed Hamiltonian with eigenstates $|\psi_1\rangle$ and $|\psi_2\rangle$ and $H'(t)$ is the time-dependent perturbation.

For the unperturbed system, the eigenvalue equation is known and obeys

$$\left\| \begin{array}{l} H_0 |\psi_1\rangle = E_1 |\psi_1\rangle \\ \\ H_0 |\psi_2\rangle = E_2 |\psi_2\rangle \end{array} \right. \tag{3.106}$$

The state of the system $|\psi(t)\rangle$ can be expanded in terms of the unperturbed states as

$$|\psi(t)\rangle = c_1(t)|\psi_1\rangle + c_2(t)|\psi_2\rangle \tag{3.107}$$

In the case of a quantum system, the time-dependent Schrödinger equation is

$$i\hbar\frac{\partial}{\partial t}|\psi(t)\rangle = H(t)|\psi(t)\rangle \tag{3.108}$$

When for this system it is applied the first-order time-dependent perturbation theory, the coefficients $c_1(t)$ and $c_2(t)$ evolve according to

$$\left| \begin{array}{l} i\hbar\frac{d}{dt}c_1(t) = E_1 c_1(t) \\ \\ i\hbar\frac{d}{dt}c_2(t) = E_2 c_2(t) + \langle\psi_2| H'(t) |\psi_1\rangle c_1(t) \end{array} \right. \tag{3.109}$$

If it is considered that $c_1(t) \approx 1$ (since initially the system is in state $|\psi_1\rangle$, and the probability is normalized in this state), we solve for $c_2(t)$. The transition amplitude from $|\psi_1\rangle$ to $|\psi_2\rangle$ is given by

$$c_2(t) = -\frac{i}{\hbar}\int_0^t \langle\psi_2| H'(t') |\psi_1\rangle e^{i\omega_{21}t'} dt' \tag{3.110}$$

where $\omega_{21} = \frac{E_2 - E_1}{\hbar}$.

The matrix element $\langle\psi_2| H'(t') |\psi_1\rangle$ is known as transition matrix and is given by

$$\langle\psi_2| H'(t') |\psi_1\rangle = \langle\psi_2| V_0 e^{-i\omega t} |\psi_1\rangle = V_{21} e^{-i\omega t} \tag{3.111}$$

where $V_{21} = \langle\psi_2| V_0 |\psi_1\rangle$ is known as transition matrix between $|\psi_1\rangle$ state to $|\psi_2\rangle$

Substituting this into the expression for $c_2(t)$ and carrying out the integral, we obtain

$$c_2(t) = -\frac{iV_{21}}{\hbar}\int_0^t e^{-i\omega t'}e^{i\omega_{21}t'}dt' = -\frac{iV_{21}}{\hbar}\int_0^t e^{i(\omega_{21}-\omega)t'}dt' \tag{3.112}$$

$$c_2(t) = -\frac{iV_{21}}{\hbar}\left[\frac{e^{i(\omega_{21}-\omega)t'}}{i(\omega_{21}-\omega)}\right]_0^t = -\frac{iV_{21}}{\hbar}\left[\frac{e^{i(\omega_{21}-\omega)t}}{i(\omega_{21}-\omega)} - \frac{1}{i(\omega_{21}-\omega)}\right] = -\frac{iV_{21}}{\hbar}\left[\frac{e^{i(\omega_{21}-\omega)t}-1}{i(\omega_{21}-\omega)}\right] \tag{3.113}$$

then

$$c_2(t) = \frac{V_{21}}{\hbar(\omega_{21}-\omega)}\left[1-e^{i(\omega_{21}-\omega)t}\right] \tag{3.114}$$

The probability of finding the system in state $|\psi_2\rangle$ at time t is given by

$$P_{1\to 2} = |c_2(t)|^2 = \left|\frac{V_{21}}{\hbar(\omega_{21}-\omega)}\left[1-e^{i(\omega_{21}-\omega)t}\right]\right|^2 \tag{3.115}$$

When the frequency of the electromagnetic wave is close to the system's natural frequency ($\omega_{21} \approx \omega_1$), it is said to be in a resonance condition.

Finally, the transition probability is obtained as

$$P_{1\to 2} = |c_2(t)|^2 \approx \left(\frac{V_{21}}{\hbar}\right)^2 \frac{4\sin^2\left(\frac{(\omega_{21}-\omega)}{2}\right)}{(\omega_{21}-\omega)^2} \tag{3.116}$$

This example illustrates how time-dependent perturbation theory can be used to calculate transition probabilities in a quantum system under a time-dependent perturbation.

3.4 Variational Method

When we have a quantum mechanical system described by the time-independent Hamiltonian $\hat{H}(x)$, which satisfies the following eigenvalue equation:

$$\hat{H}(x)\psi_i(x) = E_i\psi_i(x) \tag{3.117}$$

When Eq. 3.117 cannot be solved exactly, finding a solution that closely approximates the exact one becomes necessary. Specifically, we aim to find the eigenfunction $\psi_p(x)$ and the energy E_p, these are close to the expected experimental results. To achieve this, a trial function $\psi_p(x)$ is proposed, allowing us to obtain a value of E_p, as close as possible to the experimentally observed energy level for the system under consideration.

There are several methods, such as the one explained above, namely, the perturbation method, whose applicability depends on the magnitude of the perturbation. Another technique called the variational method involves adjusting parameters and test functions to obtain an energy value close to the expected experimental results.

Let ψ_p be a function that behaves well mathematically and satisfies the initial and/or boundary conditions of the system. Additionally, the following inequality must be fulfilled:

$$\left| \begin{array}{l} W = \dfrac{\int_{-\infty}^{\infty} \psi_p^*(x)\hat{H}(x)\psi_p(x)dx}{\int_{-\infty}^{\infty} \psi_p^*(x)\psi_p(x)dx} \geq E_{se} \\ \\ W \geq E_{se} \end{array} \right. \tag{3.118}$$

where E_{se} is the exact energy of the state under study, which in many cases may be the ground state of the system, E_0.

If the system is considered to be normalized, such that

$$\int_{-\infty}^{\infty} \psi_p^*(x)\psi_p(x)dx = 1 \tag{3.119}$$

In this case, the function $\psi_p(x)$ can be written as a linear combination of the basis functions. This is because the Hamiltonian is a Hermitian and linear operator, and the functions $\psi_i(x)$ form a complete set.

$$\left| \begin{array}{l} \psi_p(x) = \sum_i C_i \psi_i(x) \\ \\ \psi_p^*(x) = \sum_i C_i^* \psi_i^*(x) \end{array} \right. \tag{3.120}$$

Therefore, Eq. 3.118 can be written as

$$W = \int_{-\infty}^{\infty} \psi_p^*(x)\hat{H}(x)\psi_p(x)dx = \int_{-\infty}^{\infty} \left(\sum_i C_i^* \psi_i^*(x) \right) \hat{H}(x) \left(\sum_j C_j \psi_j(x) \right) dx \tag{3.121}$$

and

$$\left| \begin{array}{l} W = \int\limits_{-\infty}^{\infty} \left(\sum_i C_i^* \psi_i^*(x) \right) \left(\sum_j C_j \hat{H}(x) \psi_j(x) \right) dx \\ W = \int\limits_{-\infty}^{\infty} \left(\sum_i C_i^* \psi_i^*(x) \right) \left(\sum_j C_j E_j \psi_j(x) \right) dx \end{array} \right. \tag{3.122}$$

then

$$\left| \begin{array}{l} W = \int\limits_{-\infty}^{\infty} \left(\sum_i \sum_j C_i^* C_j E_j \psi_i^*(x) \psi_j(x) \right) dx \\ W = \sum_i \sum_j C_i^* C_j E_j \int\limits_{-\infty}^{\infty} \psi_i^*(x) \psi_j(x) dx \\ W = \sum_i \sum_j C_i^* C_j E_j \delta_{ij} \\ W = \sum_i C_i^* C_i E_i \end{array} \right. \tag{3.123}$$

And finally,

$$W = \sum_i |C_i|^2 E_i \tag{3.124}$$

But it is remembered that

$$\int\limits_{-\infty}^{\infty} \psi_i^*(x) \psi_j(x) dx = \delta_{ij}, \quad \text{and} \quad C_i^* C_i = |C_i|^2 \tag{3.125}$$

Using the normalization condition and the linear combination of the basis functions for $\psi_p(x)$ and their conjugates $\psi_p^*(x)$, the following expression is obtained:

$$\left| \begin{array}{l} 1 = \int\limits_{-\infty}^{\infty} \psi_p^*(x) \psi_p(x) dx = \int\limits_{-\infty}^{\infty} \left(\sum_i C_i^* \psi_i^*(x) \right) \left(\sum_j C_j \psi_j(x) \right) dx \\ 1 = \int\limits_{-\infty}^{\infty} \left(\sum_i \sum_j C_i^* C_j \psi_i^*(x) \psi_j(x) \right) dx = \sum_i \sum_j C_i^* C_j \int\limits_{-\infty}^{\infty} \psi_i^*(x) \psi_j(x) dx \\ 1 = \sum_i \sum_j C_i^* C_j \delta_{ij} = \sum_i |C_i|^2 \end{array} \right. \tag{3.126}$$

then

$$\sum_i |C_i|^2 = 1 \tag{3.127}$$

3.4.1 Example of the Variational Method

Let's consider the example of the harmonic oscillator [2]. In this case, the trial function can be used as follows:

$$\psi_p(x) = e^{-x^2} \tag{3.128}$$

The above function is well-behaved and satisfies the following condition:

$$\left| \begin{array}{l} x \to 0, \quad \psi_p(0) = 1 \quad \text{Finite} \\ \text{x} \to \pm\infty, \quad \lim\limits_{x \to \pm\infty} \frac{1}{e^{x^2}} \to 0 \end{array} \right. \tag{3.129}$$

Let's consider the following Hamiltonian:

$$\hat{H}(x) = -\frac{\hbar^2}{2m}\frac{d^2}{dx^2} + \frac{1}{2}kx^2 \tag{3.130}$$

And,

$$\int\limits_{-\infty}^{\infty} \psi_p^*(x)\psi_p(x)dx = \int\limits_{-\infty}^{\infty} e^{-x^2}e^{-x^2}dx = \int\limits_{-\infty}^{\infty} e^{-2x^2}dx \tag{3.131}$$

But

$$\begin{array}{l} \int\limits_0^{\infty} e^{-ax^2}dx = \frac{1}{2}\left(\frac{\pi}{a}\right)^{\frac{1}{2}}, \quad a > 0 \quad \to \quad \int\limits_{-\infty}^{\infty} e^{-ax^2}dx = 2\times\frac{1}{2}\left(\frac{\pi}{a}\right)^{\frac{1}{2}}, \quad a > 0 \\ \int\limits_{-\infty}^{\infty} e^{-ax^2}dx = \left(\frac{\pi}{a}\right)^{\frac{1}{2}}, \quad a > 0 \quad \to \forall a = 2, \quad \int\limits_{-\infty}^{\infty} e^{-2x^2}dx = \left(\frac{\pi}{2}\right)^{\frac{1}{2}} \end{array} \tag{3.132}$$

And by using the variational integral:

$$W = \frac{\int\limits_{-\infty}^{\infty} \psi_p^*(x)\hat{H}(x)\psi_p(x)dx}{\int\limits_{-\infty}^{\infty} \psi_p^*(x)\psi_p(x)dx} = \frac{\int\limits_{-\infty}^{\infty} e^{-x^2}\hat{H}(x)e^{-x^2}dx}{\int\limits_{-\infty}^{\infty} e^{-2x^2}dx} \tag{3.133}$$

then

$$\begin{aligned}
W &= \left(\tfrac{2}{\pi}\right)^{\frac{1}{2}} \int_{-\infty}^{\infty} \psi_p^*(x)\hat{H}(x)\psi_p(x)dx \\
W &= \left(\tfrac{2}{\pi}\right)^{\frac{1}{2}} \int_{-\infty}^{\infty} e^{-x^2}\left(-\tfrac{\hbar^2}{2m}\tfrac{d^2}{dx^2} + \tfrac{1}{2}kx^2\right)e^{-x^2}dx \\
W &= \left(\tfrac{2}{\pi}\right)^{\frac{1}{2}} \int_{-\infty}^{\infty} e^{-x^2}\left(-\tfrac{\hbar^2}{2m}\tfrac{d^2}{dx^2}\left(e^{-x^2}\right) + \tfrac{1}{2}kx^2e^{-x^2}\right)dx
\end{aligned} \tag{3.134}$$

But

$$-\frac{\hbar^2}{2m}\frac{d^2}{dx^2}\left(e^{-x^2}\right) = -\frac{\hbar^2}{2m}\left[-2e^{-x^2} + 4x^2e^{-x^2}\right] \tag{3.135}$$

Therefore,

$$\begin{aligned}
W &= \left(\tfrac{2}{\pi}\right)^{\frac{1}{2}} \int_{-\infty}^{\infty} e^{-x^2}\left[-\tfrac{\hbar^2}{2m}\left(-2e^{-x^2} + 4x^2e^{-x^2}\right) + \tfrac{1}{2}kx^2e^{-x^2}\right]dx \\
W &= \left(\tfrac{2}{\pi}\right)^{\frac{1}{2}}\left[\int_{-\infty}^{\infty} \tfrac{\hbar^2}{m}e^{-2x^2}dx - \int_{-\infty}^{\infty} \tfrac{2\hbar^2}{m}x^2e^{-2x^2}dx + \int_{-\infty}^{\infty} \tfrac{1}{2}kx^2e^{-2x^2}dx\right]
\end{aligned} \tag{3.136}$$

Taking into account that:

$$\begin{aligned}
&\int_0^{\infty} x^{2n}e^{-ax^2}dx = \tfrac{1\cdot 2\cdot 3\ldots(2n-1)}{2^{n+1}}\left(\tfrac{\pi}{a^{2n+1}}\right)^{\frac{1}{2}}, \quad a > 0, \quad \text{n} = 1, 2, 3\ldots \\
&\int_{-\infty}^{\infty} x^{2n}e^{-ax^2}dx = 2 \times \int_0^{\infty} x^{2n}e^{-ax^2}dx = \tfrac{1\cdot 2\cdot 3\ldots(2n-1)}{2^{n}}\left(\tfrac{\pi}{a^{2n+1}}\right)^{\frac{1}{2}}, \quad a > 0, \quad \text{n} = 1, 2, 3\ldots
\end{aligned} \tag{3.137}$$

It is obtained

$$\begin{aligned}
&\int_{-\infty}^{\infty} \tfrac{2\hbar^2}{m}x^2e^{-2x^2}dx = \tfrac{2\hbar^2}{m}\tfrac{1}{2}\left(\tfrac{\pi}{2^3}\right)^{\frac{1}{2}} = \tfrac{\hbar^2}{2m}\left(\tfrac{\pi}{2}\right)^{\frac{1}{2}} \\
&\int_{-\infty}^{\infty} \tfrac{\hbar^2}{m}e^{-2x^2}dx = \tfrac{\hbar^2}{m}\left(\tfrac{\pi}{2}\right)^{\frac{1}{2}} \\
&\int_{-\infty}^{\infty} \tfrac{1}{2}kx^2e^{-2x^2}dx = \tfrac{1}{2}k\tfrac{1}{2}\left(\tfrac{\pi}{2^3}\right)^{\frac{1}{2}} = \tfrac{1}{8}k\left(\tfrac{\pi}{2}\right)^{\frac{1}{2}}
\end{aligned} \tag{3.138}$$

Finally, it is obtained for variational integral

$$\begin{aligned} &W = \left(\tfrac{2}{\pi}\right)^{\frac{1}{2}} \left[\tfrac{\hbar^2}{m}\left(\tfrac{\pi}{2}\right)^{\frac{1}{2}} - \tfrac{\hbar^2}{2m}\left(\tfrac{\pi}{2}\right)^{\frac{1}{2}} + \tfrac{1}{8}k\left(\tfrac{\pi}{2}\right)^{\frac{1}{2}}\right] \\ &W = \left(\tfrac{2}{\pi}\right)^{\frac{1}{2}}\left(\tfrac{\pi}{2}\right)^{\frac{1}{2}} \left[\tfrac{\hbar^2}{m} - \tfrac{\hbar^2}{2m} + \tfrac{1}{8}k\right] \\ &W = \tfrac{\hbar^2}{2m} + \tfrac{1}{8}k \end{aligned} \tag{3.139}$$

The exact energy for the harmonic oscillator is

$$E_{Oscillator} = \left(n + \frac{1}{2}\right)\hbar\omega, \quad \text{with} \quad \omega = \sqrt{\frac{k}{m}} \tag{3.140}$$

$$E_{Oscillator} = \frac{1}{2}\hbar\omega = \frac{1}{2}\hbar\sqrt{\frac{k}{m}} \tag{3.141}$$

Taking into account the condition for variational integral as

$$W \geq E_{Oscillator} \quad \rightarrow \quad W - E_{Oscillator} \geq 0 \tag{3.142}$$

After substituting and simplifying, we get

$$\begin{aligned} &\tfrac{\hbar^2}{2m} + \tfrac{1}{8}k \geq \tfrac{1}{2}\hbar\sqrt{\tfrac{k}{m}} \\ &\tfrac{1}{2}\hbar\sqrt{\tfrac{k}{m}}\left(\tfrac{\hbar}{m}\sqrt{\tfrac{m}{k}}\right) + \tfrac{1}{2}\hbar\sqrt{\tfrac{k}{m}}\left(\tfrac{m}{4\hbar}\sqrt{\tfrac{k}{m}}\right) \geq \tfrac{1}{2}\hbar\sqrt{\tfrac{k}{m}} \\ &\left(\tfrac{\hbar}{m}\sqrt{\tfrac{m}{k}}\right) + \left(\tfrac{m}{4\hbar}\sqrt{\tfrac{k}{m}}\right) \geq 1 \end{aligned} \tag{3.143}$$

By using the following algebraic relation:

$$\begin{aligned} &\tfrac{a}{b} + \tfrac{b}{4a} \geq 1 \\ &x = \tfrac{a}{b} = \tfrac{\hbar}{\sqrt{km}} \\ &x + \tfrac{1}{4x} \geq 1 \\ &4x^2 + 1 \geq 4x \\ &4x^2 - 4x + 1 \geq 0 \end{aligned} \tag{3.144}$$

A graphical solution is shown for the above equation (Fig. 3.1), then,

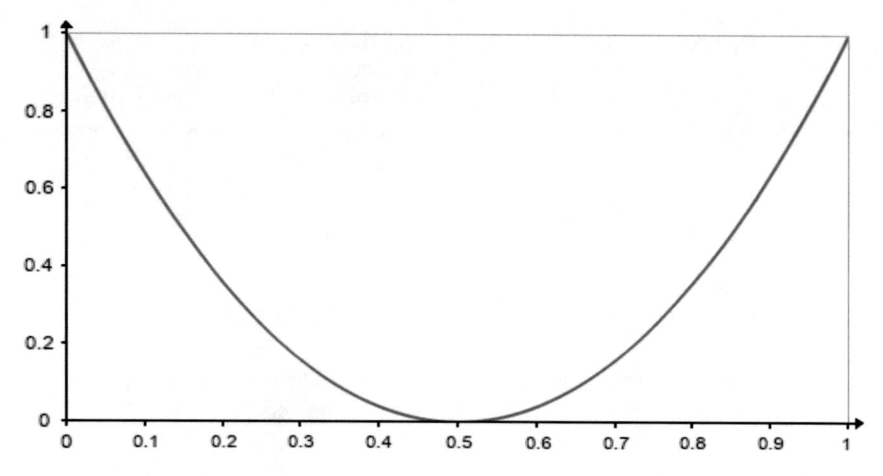

Fig. 3.1 Graphical Solution

$$\left| \begin{array}{l} x = 0.50 = \frac{1}{2} = \frac{\hbar}{\sqrt{km}} \\ \frac{\sqrt{km}}{2\hbar} = 1 \end{array} \right. \tag{3.145}$$

The above equation indicates that for the test function to be well-behaved, one must ensure that the relationship between k and m is restricted to the values of equation x. However, this would not be the proper approach, as the physical phenomenon should not be constrained by the mathematical results. In such a case, we must reconsider the test function used.

When Eq. 3.145 is substituted into 3.139, the result is obtained as

$$W = \frac{\hbar^2}{2m} + \frac{1}{8}k = \frac{\hbar^2}{2m}\left[\frac{\sqrt{km}}{2\hbar}\right] + \frac{1}{8}k\left[\frac{2\hbar}{\sqrt{km}}\right] \tag{3.146}$$

In above equation uses

$$\frac{\sqrt{km}}{2\hbar} = 1, \quad \rightarrow \quad \frac{2\hbar}{\sqrt{km}} = 1 \tag{3.147}$$

Therefore,

$$\left| \begin{array}{l} W = \frac{\hbar^2}{2m} + \frac{1}{8}k = \frac{\hbar^2}{2m}\left[\frac{\sqrt{km}}{2\hbar}\right] + \frac{1}{8}k\left[\frac{2\hbar}{\sqrt{km}}\right] \\ W = \frac{\hbar}{2}\sqrt{\frac{k}{m}} + \frac{\hbar}{4}\sqrt{\frac{k}{m}} = \frac{3\hbar}{4}\sqrt{\frac{k}{m}} \end{array} \right. \tag{3.148}$$

It is observed that the energy W is greater than the fundamental level above $\frac{\hbar}{4}\sqrt{\frac{k}{m}}$, and this result lies between $\frac{\hbar}{2}\sqrt{\frac{k}{m}}$ and $\frac{3\hbar}{2}\sqrt{\frac{k}{m}}$, as is observed in equation. Therefore, it is concluded that the function used to find the solution to the problem is not optimal.

$$\left| \begin{array}{l} E_{Oscillator}(0) = \left(0 + \frac{1}{2}\right)\hbar\sqrt{\frac{k}{m}} = \frac{\hbar}{2}\sqrt{\frac{k}{m}}, \quad \text{ground} - \text{state} \\ \\ E_{Oscillator}(1) = \left(1 + \frac{1}{2}\right)\hbar\sqrt{\frac{k}{m}} = \frac{3\hbar}{2}\sqrt{\frac{k}{m}}, \quad \text{first} - \text{excited} - \text{state} \end{array} \right. \tag{3.149}$$

One way to ensure that the test function is appropriate is to introduce parameters α into the function, such that by adjusting these parameters, a function is obtained that closely approximates the exact energy of the system, which is the goal of the calculation. For example, it is left as an exercise to obtain the energy using the variational method and to try the following function:

$$\psi_p(x) = e^{-\alpha x^2}, \forall \alpha > 0 \tag{3.150}$$

The function has the condition that

$$\left| \begin{array}{l} x \to 0, \quad \psi_p(0) = 1 \quad \text{Finite} \\ \text{x} \to \pm\infty, \quad \lim\limits_{x \to \pm\infty} \frac{1}{e^{\alpha x^2}} \to 0 \end{array} \right. \tag{3.151}$$

Using 3.118, and taking account that

$$\int_{-\infty}^{\infty} \psi_p^*(x)\psi_p(x)dx = \int_{-\infty}^{\infty} e^{-\alpha x^2}e^{-\alpha x^2}dx = \int_{-\infty}^{\infty} e^{-2\alpha x^2}dx \tag{3.152}$$

Using 3.132, it is obtained that

$$\int_{-\infty}^{\infty} \psi_p^*(x)\psi_p(x)dx = \left(\frac{\pi}{2\alpha}\right)^{\frac{1}{2}}, \quad \forall \alpha > 0 \tag{3.153}$$

Using 3.118, it is obtained for W,

$$W = \left(\frac{\pi}{2\alpha}\right)^{-\frac{1}{2}} \int_{-\infty}^{\infty} \psi_p^*(x)\hat{H}(x)\psi_p(x)dx, \quad \text{with } \hat{H}(x) = -\frac{\hbar^2}{2m}\frac{d^2}{dx^2} + \frac{1}{2}kx^2 \tag{3.154}$$

then

$$\begin{aligned} W &= \left(\tfrac{\pi}{2\alpha}\right)^{-\frac{1}{2}} \int_{-\infty}^{\infty} e^{-\alpha x^2}\left(-\tfrac{\hbar^2}{2m}\tfrac{d^2}{dx^2} + \tfrac{1}{2}kx^2\right) e^{-\alpha x^2} dx \\ W &= \left(\tfrac{\pi}{2\alpha}\right)^{-\frac{1}{2}} \int_{-\infty}^{\infty} e^{-\alpha x^2}\left(-\tfrac{\hbar^2}{2m}\tfrac{d^2}{dx^2}\left(e^{-\alpha x^2}\right) + \tfrac{1}{2}kx^2 e^{-\alpha x^2}\right) dx \end{aligned} \tag{3.155}$$

Taking account that

$$\begin{aligned} \hat{H}(x)\psi_p(x) &= -\tfrac{\hbar^2}{2m}\tfrac{d^2}{dx^2}\left(e^{-\alpha x^2}\right) + \tfrac{1}{2}kx^2 e^{-\alpha x^2} \\ &= -\tfrac{\hbar^2}{2m}\left(-2\alpha e^{-\alpha x^2} + 4\alpha^2 x^2 e^{-\alpha x^2}\right) + \tfrac{1}{2}kx^2 e^{-\alpha x^2} \end{aligned} \tag{3.156}$$

Substituting in (3.156), we obtain

$$W = \left(\frac{\pi}{2\alpha}\right)^{-\frac{1}{2}} \left[\frac{\hbar^2\alpha}{m} \int_{-\infty}^{\infty} e^{-2\alpha x^2} dx - \left(\frac{2\hbar^2\alpha^2}{m} - \frac{1}{2}k\right) \int_{-\infty}^{\infty} x^2 e^{-2\alpha x^2} dx\right] \tag{3.157}$$

and using the integrals given for (3.132) and (3.137), we obtain

$$W = \left(\frac{\pi}{2\alpha}\right)^{-\frac{1}{2}} \left[\frac{\hbar^2\alpha}{m}\left(\frac{\pi}{2\alpha}\right)^{\frac{1}{2}} - \left(\frac{2\hbar^2\alpha^2}{m} - \frac{1}{2}k\right)\frac{1}{2}\left(\frac{\pi}{(2\alpha)^3}\right)^{\frac{1}{2}}\right] \tag{3.158}$$

finally, simplifying it, we obtain

$$W(\alpha) = \frac{\hbar^2\alpha}{2m} + \frac{1}{8}\frac{k}{\alpha} \tag{3.159}$$

When the quantum oscillator was considered, the energies were given by the following equation:

$$E_{Oscillator} = \left(n + \frac{1}{2}\right)\hbar\sqrt{\frac{k}{m}} \tag{3.160}$$

The energies obtained were exact. In this context, the energy determined by the variational method depends on the parameter α. It must differentiate the function W concerning α to minimize the expression $W(\alpha)$. Doing this makes it possible to minimize $W(\alpha)$ and find the optimal value of α. This value is then substituted into Eq. 3.159. As shown, α helps to obtain the appropriate function for applying the variational method, resulting in an energy equal to the value found in the previous section.

$$\frac{dW(\alpha)}{dx} = \frac{\hbar^2}{2m} - \frac{1}{8}\frac{k}{\alpha^2} = 0, \quad \alpha = \pm\frac{\sqrt{km}}{2\hbar}, \text{ but } \alpha \geq 0, \text{ then } \alpha = \frac{\sqrt{km}}{2\hbar} \tag{3.161}$$

Therefore, substituting the α value in (3.159) equation has

$$W = \frac{\hbar^2 \frac{\sqrt{km}}{2\hbar}}{2m} + \frac{1}{8}k\left(\frac{2\hbar}{\sqrt{km}}\right) = \frac{\hbar}{4}\sqrt{\frac{k}{m}} + \frac{\hbar}{4}\sqrt{\frac{k}{m}} = \frac{\hbar}{2}\sqrt{\frac{k}{m}} \tag{3.162}$$

This is the exact energy for the harmonic oscillator at the fundamental level, i.e., n = 0.

3.5 Proposed Problems

3.1 Calculate the first state of the harmonic oscillator when the perturbation has a functional expression of the form $V(x) = \alpha x^3$.

3.2 Calculate the first state of the harmonic oscillator when the perturbation has a functional expression of the form $V(x) = \beta x$.

3.3 Consider a hydrogen atom with an electron in the ground state. A small perturbation of an electric field E is applied along the z-axis, resulting in a perturbation potential of the form $V = eEz$, where e is the electron charge and z is the position coordinate along the z-axis.

- Calculate the first-order energy correction to the ground state due to the perturbation.
- Conclude whether the first-order energy correction to the ground state is zero or non-zero, and explain why it does or does not occur.

3.4 Consider a two-level quantum system with energy eigenstates $|1\rangle$ and $|2\rangle$ with corresponding energies E_1 and E_2. The system is initially in the state $|1\rangle$ at time t = 0.

A time-dependent perturbation is applied to the system, which is given by

$$V(t) = V_0 \cos(\omega t) \left[|1\rangle \langle 2| + |2\rangle \langle 1|\right] \tag{3.163}$$

where V_0 is a constant and ω is the frequency of the perturbation.

- Write down the time-dependent Schrödinger equation for this system.
- Calculate the first-order transition amplitude $c_2(t)$ from the initial state $|1\rangle$ to the final state $|2\rangle$ at time t.
- Obtain the transition probability $P_{|1\rangle \to |2\rangle}$ that the system is in state $|2\rangle$ at time t.

- Analyze and discuss how the transition probability $P_{|1\rangle \to |2\rangle}$ depends on the parameters V_0, ω, and the time t, and what happens if these parameters change. Explain how resonance conditions affect the transition probability and how it is modified if the frequency of the perturbation is changed.

3.5 Consider a quantum harmonic oscillator with the Hamiltonian:

$$H_0 = \hbar\omega\left(a^+ a + \frac{1}{2}\right) \tag{3.164}$$

where a^+ and a are the creation and annihilation operators, and ω is the angular frequency of the oscillator. A time-dependent perturbation is applied to the system, given by

$$V(t) = \lambda \hat{x} \cos(\omega t) \tag{3.165}$$

where λ is a small constant and $\hat{x}$ is the position operator, which is given by

$$\hat{x} = \sqrt{\frac{\hbar}{2m\omega}(a^+ + a)} \tag{3.166}$$

Assume the system is initially in the ground state $|0\rangle$ at time t = 0. Calculate:

- Write down the time-dependent Schrödinger equation for this system.
- Calculate the first-order transition amplitude $c_1(t)$ from the initial state $|0\rangle$ to the first excited state $|1\rangle$ at time t.
- Compute the transition probability $P_{|0\rangle \to |1\rangle}(t)$ to find the probability that the system is in state $|1\rangle$ at time t.
- Analyze and discuss how the transition probability $P_{|0\rangle \to |1\rangle}(t)$ depends on the parameters λ, ω, and the time t, and what happens if these parameters change. Explain how resonance conditions affect the transition probability and how it is modified if the frequency of the perturbation is changed.

3.6 Using the $\psi_p(x) = e^{-\alpha\hbar x^2}$ function shown below, obtain the energy of the harmonic oscillator at its fundamental level, E_0. It is recommended to use the variational method.

References

1. Levine IN (2014) Quantum mechanics, 7th ed. Pearson
2. Requena Rodriguez A, Zuñiga Roman J (2004) Espectroscopia, pearson educacion, SA, Madrid

Chapter 4
Harmonic Crystals

In this book, the harmonic crystal will be defined based on its behavior, which has a periodic distribution in space and follows sine and cosine functions for its physical explanation. From the standpoint of periodicity, order, and regular behavior, its response can be described as a harmonic function, which is a linear combination of sine and cosine functions.

When looking for a solution to a physical system or problem from the point of view of classical mechanics, Newton's equations are used as the foundation of the classical explanation. Often, using Newton's equations in their original form produces mathematical difficulties. Therefore, in most cases of classical mechanics, problems are addressed using Lagrange's equations, which allow solving the problem through the respective differential equations. Below, examples of the application of Lagrange's equations to obtain solutions for vibratory and rotational systems will be shown [1, 2].

4.1 Lagrangian Equation

Classical mechanics is founded on Newton's laws, and in many dynamics problems, the second law expressed as $\mathbf{F} = m\mathbf{a}$ is essential. However, there are methods that provide greater effectiveness and more generalized views, among which are two classical methods: the Hamilton principle and the Lagrange equations.

The Lagrangian function or Lagrangian is defined as $L = T - V$, where T is the kinetic energy and V the potential energy that depends on the generalized coordinates q and the generalized velocities $\dot{q}$, that is; $T(q, \dot{q})$ and $V(q, \dot{q})$. However, in a *conservative system, the potential energy depends only on the generalized coordinates;* $V(q)$, the Lagrange equation is written as

C. Vargas Hernandez, *Introduction to Raman Spectroscopy and Its Applications*,
https://doi.org/10.1007/978-3-031-77551-2_4

$$\frac{d}{dt}\left[\frac{\partial L}{\partial \dot{q}_k}\right] - \frac{\partial L}{\partial q_k} = Q_k \tag{4.1}$$

The generalized coordinates q_k are a set of coordinates that must uniquely define the state of the system. Phenomenologically, they are linearly independent coordinates, or in other words, the minimum number of generalized coordinates to define the state of the system is known as: independent coordinates. Thus, the set of q_k must equal the degrees of freedom of the system under study. The Q_k are known as generalized forces and $k = 1, 2, 3, ..., n$. In the case of a conservative system, the force derives from a potential, that is $Q_k = -\frac{\partial V}{\partial q_k}$, and therefore the equation can be written as

$$\frac{d}{dt}\left[\frac{\partial T}{\partial \dot{q}_k}\right] - \frac{\partial T}{\partial q_k} = Q_k \tag{4.2}$$

Although not proven in this text, this equation is also correct for non-conservative systems, with the only difference being that in a non-conservative system, the generalized force cannot be obtained from a potential function, meaning it does not derive from a potential, such that $Q_k = -\frac{\partial V}{\partial q_k}$.

4.2 Normal Coordinates

Normal coordinates are a set of generalized coordinates that decouple the differential equations. The systems of differential equations for solving problems in coupled oscillations, as occurs in molecules, can be solved by changing variables leading to the so-called normal coordinates or the amplitudes of molecular vibration modes, or normal vibration modes. These normal coordinates are a particular set of generalized coordinates to solve the initial mechanical problem. Coupled oscillations are a model solution for molecules present in a crystal and help explain thermal vibrations in a crystal [3, 4].

A crystalline material is defined as a material where atoms occupy regular positions that repeat periodically, forming an ideal lattice or periodic grid. When these atoms vibrate around equilibrium points, explained through periodic oscillations and expressed by sine or cosine functions called harmonic functions, we are talking about **Harmonic Crystals**.

Example Consider a mass-spring system with a mass m and a spring with a recovery constant k. This model is applicable to atoms vibrating around centers that displace minimally, or molecules where one atom has a much greater mass than the other. Consider the figure below, where the mass m is subjected to a recovery force obeying Hooke's law (Fig. 4.1):

$$\mathrm{F} = -\mathrm{Kx}$$

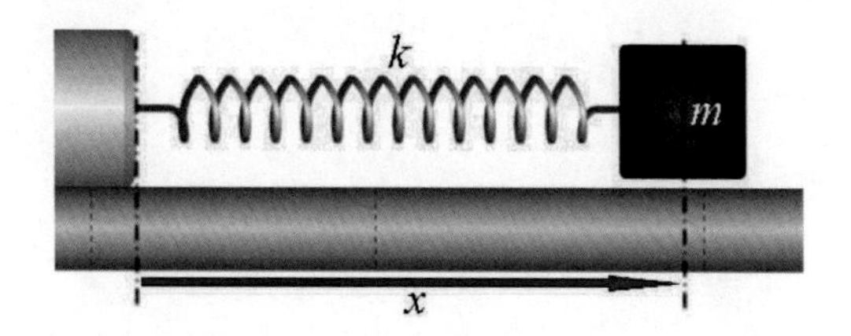

Fig. 4.1 Mass-spring system

Applying the Eq. 4.2,

$$\frac{\partial}{\partial t}\left(\frac{\partial T}{\partial \dot{q}_k}\right) - \frac{\partial T}{\partial q_k} = Q_k \tag{4.3}$$

with

$$\bar{F} = -\bar{\nabla}_{q_k} V \tag{4.4}$$

$$\Rightarrow \frac{\partial}{\partial t}\left(\frac{\partial L}{\partial \dot{q}_k}\right) - \frac{\partial L}{\partial q_k} = 0 \tag{4.5}$$

$$L = T - L \tag{4.6}$$

From the Eq. 4.218, it is obtained

$$V = \int F.dx = \frac{1}{2}Kx^2 \tag{4.7}$$

Therefore,

$$L = \frac{1}{2}m\dot{x}^2 - \frac{1}{2}kx^2 \tag{4.8}$$

So, using:

$$\frac{\partial}{\partial t}\left(\frac{\partial L}{\partial \dot{q}_k}\right) - \frac{\partial L}{\partial q_k} = 0 \tag{4.9}$$

with $q_k = x$

$$\frac{\partial L}{\partial x} = -kx, \qquad \frac{\partial L}{\partial \dot{x}} = m\dot{x}, \qquad \frac{\partial}{\partial t}\left(\frac{\partial L}{\partial \dot{x}}\right) = m\ddot{x} \tag{4.10}$$

$$m\ddot{x} + kx = 0 \tag{4.11}$$

$$x = x_0 e^{i\omega t}, \qquad \dot{x} = x_0 i\omega e^{i\omega t}, \qquad \ddot{x} = -x_0\omega^2 e^{i\omega t} \tag{4.12}$$

$$-x_o m\omega^2 e^{i\omega t} + kx_o e^{i\omega t} = 0 \quad \Rightarrow \quad k = m\omega^2 \tag{4.13}$$

The natural frequency of the system is obtained as

$$w_o = \sqrt{\frac{k}{m}} \tag{4.14}$$

Consider Fig. 4.17 which has three masses m_1, m_2 and m_3 connected by four springs with spring constants k_1, k_{12}, k_{23} and k_3, with fixed ends at their extremities, and initial conditions for the positions and velocities: $x_{10}, x_{20}, x_{30}, v_{10}, v_{20}, v_{30}$. Coordinates $x_1, x_2 - x_1, x_3 - x_2, -x_3$. Due to the existence of four springs, there are four forces acting on the system, with friction between the masses and the horizontal surface neglected. For this proposed exercise, find

(a) The normal frequencies of vibration.
(b) The normal coordinates of vibration.

Example
Consider the Fig. 4.18, with three equal masses m connected by two springs with equal spring constants k, without friction between the masses and the horizontal surface (Fig. 4.2).

Therefore,

$$L = T - L \tag{4.15}$$

Such that

$$T = \frac{1}{2}m\dot{x}_1^2 + \frac{1}{2}m\dot{x}_2^2 + \frac{1}{2}m\dot{x}_3^2 \tag{4.16}$$

$$V = \frac{1}{2}k(x_2 - x_1)^2 + \frac{1}{2}k(x_3 - x_2)^2 \tag{4.17}$$

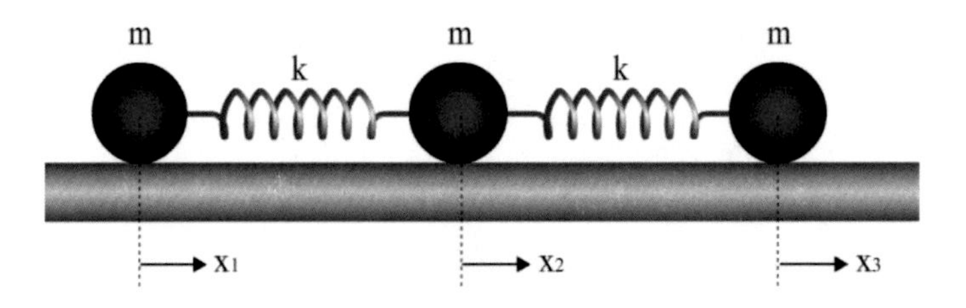

Fig. 4.2 System of 3 coupled masses

Then,

$$L = \frac{1}{2}m\dot{x}_1^2 + \frac{1}{2}m\dot{x}_2^2 + \frac{1}{2}m\dot{x}_3^2 - \frac{1}{2}k(x_2 - x_1)^2 - \frac{1}{2}k(x_3 - x_2)^2 \tag{4.18}$$

$$\frac{\partial}{\partial t}\left(\frac{\partial L}{\partial \dot{q}_k}\right) - \frac{\partial L}{\partial q_k} = 0 \qquad [q_k = x_1, x_2, x_3] \tag{4.19}$$

Since there are three generalized coordinates, this implies three differential equations.

$$\frac{\partial L}{\partial q_k} = \frac{\partial L}{\partial x_1}, \frac{\partial L}{\partial x_2}, \frac{\partial L}{\partial x_3} \tag{4.20}$$

$$\frac{\partial L}{\partial x_1} = -k(x_2 - x_1) \qquad \frac{\partial L}{\partial \dot{x}_1} = m\dot{x}_1 \tag{4.21}$$

$$\begin{aligned} m\ddot{x}_1 - k\,(x_2 - x_1) = 0 \qquad & (a) \\ m\ddot{x}_2 + k\,(x_2 - x_1) - k\,(x_3 - x_2) = 0 \; & (b) \\ m\ddot{x}_3 + k\,(x_3 - x_2) = 0 \qquad & (c) \end{aligned} \tag{4.22}$$

Let

$$x_j = x_{0j}e^{i\omega t}, \quad \dot{x}_j = i\omega x_j, \quad \ddot{x}_j = i\omega \dot{x}_j, \quad \ddot{x}_j = -\omega^2 x_j \tag{4.23}$$

where $x_{0j} = a_j$

Then, substituting

$$\begin{aligned} ma_1(-\omega^2) + ka_1 - ka_2 = 0 \\ ma_2(-\omega^2) - ka_1 + 2ka_2 - ka_3 = 0 \\ ma_3(-\omega^2) - ka_2 + ka_3 = 0 \end{aligned}$$

Let

$$\lambda = \frac{\omega^2 m}{k} \tag{4.24}$$

$$\begin{aligned} (1) \qquad -a_1\lambda + a_1 - a_2 = 0 \\ (2)\; -a_2\lambda - a_1 + 2a_2 - a_3 = 0 \\ (3) \qquad -a_3\lambda - a_2 + a_3 = 0 \end{aligned} \tag{4.25}$$

The previous expression is a secular equation, defined by

$$MA = \lambda A \tag{4.26}$$

$$MA - \lambda A = 0 \tag{4.27}$$

$$(M - \lambda)A = 0 \tag{4.28}$$

$$A = \begin{pmatrix} a_1 \\ a_2 \\ a_3 \end{pmatrix} \tag{4.29}$$

$$M = \begin{pmatrix} 1 & -1 & 0 \\ -1 & 2 & -1 \\ 0 & 1 & 1 \end{pmatrix} \tag{4.30}$$

$$\begin{vmatrix} 1-\lambda & -1 & 0 \\ -1 & 2-\lambda & -1 \\ 0 & 1 & 1-\lambda \end{vmatrix} = 0 \tag{4.31}$$

The roots are: $\lambda_1 = 0, \quad \lambda_2 = 1, \quad \lambda_3 = 3$ According to Eq. 4.24, it is obtained $\omega_1 = 0, \quad \omega_2 = \sqrt{\frac{k}{m}}, \quad \omega_3 = \sqrt{\frac{3k}{m}}$.

Substituting into Eq. 4.25, the coordinates that define the normal modes of vibration for each value of λ are obtained as follows:

$$\lambda_1 = 0 \qquad a_1 = a_2 = a_3 \tag{4.32}$$

$$\lambda_2 = 1 \qquad a_1 = -a_3, \quad a_2 = 0 \tag{4.33}$$

$$\lambda_3 = 3 \qquad a_1 = a_3, \quad a_2 = -2a_1 \tag{4.34}$$

Using the normalization condition $\rightarrow a_i^2 = 1$, the normalization is given by $\rightarrow a_1^2 + a_2^2 + a_3^2 = 1$

$$\lambda_1 = 0 \qquad a_1 = a_2 = a_3 = \frac{1}{\sqrt{3}} \tag{4.35}$$

$$\lambda_2 = 1 \qquad a_1 = -a_3 = \frac{1}{\sqrt{2}}, \qquad a_2 = 0 \tag{4.36}$$

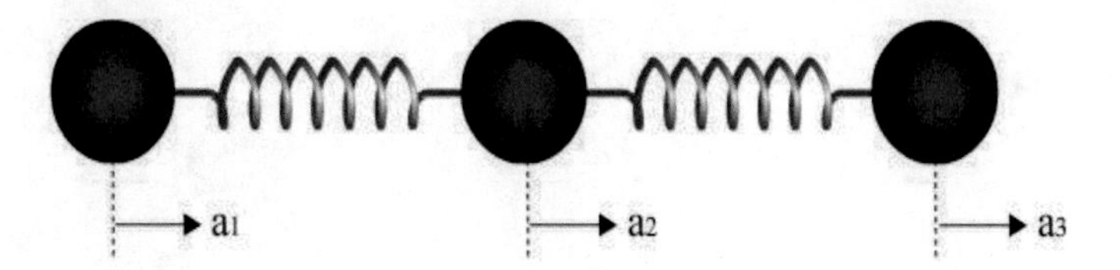

Fig. 4.3 Vibration of $\lambda_1 = 0$ (Translation)

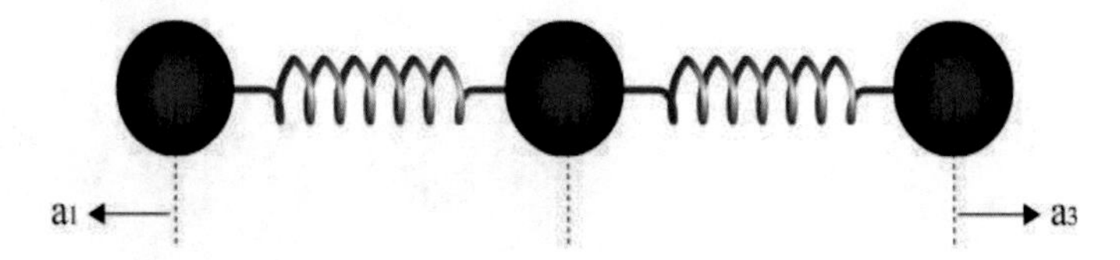

Fig. 4.4 Vibration of $\lambda_2 = 1$

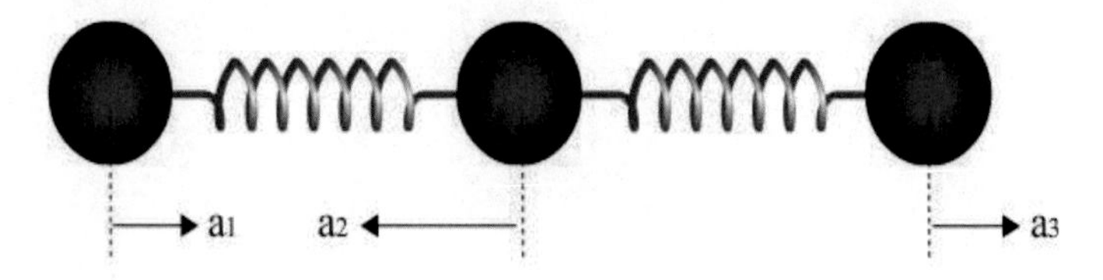

Fig. 4.5 Vibration of $\lambda_3 = 3$

$$\lambda_3 = 3 \quad a_1 = a_3 = \frac{1}{\sqrt{6}}, \quad a_2 = \frac{-2}{\sqrt{6}} \tag{4.37}$$

For $\lambda_1 = 0$, the three masses move synchronously as a whole, indicating a zero frequency $\omega = 0$, meaning it is not a vibration but a translation of the system. Figures 4.3, 4.4, and 4.5 illustrate the relative movements of the masses.

These vibration modes of the linear molecule formed by three masses represent the normal modes of vibration of the molecule subjected to the constraints described in the example and are the fingerprint characterizing this system.

4.3 Planar Equilateral Triangle

To illustrate, we will solve for a molecule formed by three atoms located at the corners of a planar equilateral triangle, see Fig. 4.6. This example is a prototype of a nonlinear molecule and is a useful model as this type of symmetry exists in molecular structures in nature.

It is assumed that the masses m of the atoms are equal and the springs that couple them have the same spring constant k. The goal is to answer the following questions:

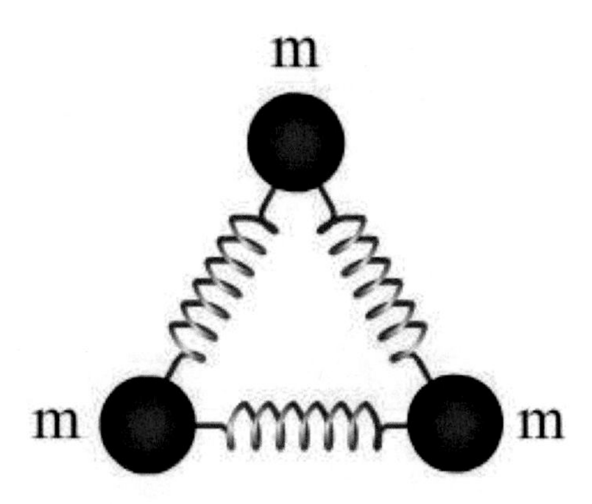

Fig. 4.6 System of 3 masses coupled in a planar equilateral triangle

(a) Obtain the normal frequencies of vibration.
(b) Find the normal coordinates of vibration.

To do this, we must find the Lagrangian of the system, i.e., we seek

$$L = T - V \tag{4.38}$$

The kinetic energy of the system is

$$T = K_{1x} + K_{1y} + K_{2x} + K_{2y} + K_{3x} + K_{3y} \tag{4.39}$$

$$T = \frac{m_1}{2}(\dot{x}_1)^2 + \frac{m_2}{2}(\dot{y}_1)^2 + \frac{m_2}{2}(\dot{x}_2)^2 + \frac{m_2}{2}(\dot{y}_2)^2 + \frac{m_3}{2}(\dot{x}_3)^2 + \frac{m_3}{2}(\dot{y}_3)^2 \tag{4.40}$$

Since the atoms are identical, $m_1 = m_2 = m_3 = m$, therefore the total kinetic energy T is given by the expression:

$$T = \frac{m}{2}\left[\dot{x}_1^2 + \dot{y}_1^2 + \dot{x}_2^2 + \dot{y}_2^2 + \dot{x}_3^2 + \dot{y}_3^2\right] \tag{4.41}$$

The potential energy V for each spring is given by the functional expression:

$$V = \frac{1}{2}kx^2 \tag{4.42}$$

Therefore, the total potential energy of the system is

$$V_T = V_1 + V_2 + V_3 \tag{4.43}$$

Since the atoms are identical, $k_1 = k_2 = k$, so the total potential energy is given by

$$V_T = \frac{k}{2}\left\{(x_3 - x_2)^2 \left[-\frac{1}{2}(x_1 - x_3) + \frac{\sqrt{3}}{2}(y_1 - y_3)\right]^2 + \left[\frac{1}{2}(x_2 - x_1) + \frac{\sqrt{3}}{2}(y_2 - y_1)\right]^2\right\} \tag{4.44}$$

$$V_T = \frac{k}{2}\left\{\begin{array}{l}(x_3^2 - 2x_3x_2 + x_2^2) + \left[\frac{1}{4}(x_1 - x_3)^2 - \frac{\sqrt{3}}{2}(x_1 - x_3)(y_1 - y_3) + \frac{3}{4}(y_1 - y_3)^2\right] + \\ \left[\frac{1}{4}(x_2 - x_1)^2 - \frac{\sqrt{3}}{2}(x_2 - x_1)(y_2 - y_1) + \frac{3}{4}(y_2 - y_1)^2\right]\end{array}\right\} \tag{4.45}$$

$$V_T = \frac{k}{2}\left\{\begin{array}{l}(x_3^2 - 2x_3x_2 + x_2^2) + \left[\begin{array}{l}\frac{1}{4}(x_1^2 - 2x_1x_3 - x_3^2) - \frac{\sqrt{3}}{2}(x_1y_1 - x_1y_3 - x_3y_1 + x_3y_3) \\ +\frac{3}{4}(y_1^2 - 2y_1y_3 + y_3^2)\end{array}\right] + \\ \left[\begin{array}{l}\frac{1}{4}(x_2^2 - 2x_2x_1 + x_1^2) - \frac{\sqrt{3}}{2}(x_2y_2 - x_2y_1 - x_1y_2 + x_1y_1) \\ +\frac{3}{4}(y_2^2 - 2y_2y_1 + y_1^2)\end{array}\right]\end{array}\right\} \tag{4.46}$$

$$V_T = \frac{k}{2}\left\{\begin{array}{l}x_3^2 - 2x_3x_2 + x_2^2 + \frac{1}{4}x_1^2 - \frac{1}{2}x_1x_3 + \frac{1}{4}x_3^2 - \frac{\sqrt{3}}{2}x_1y_1 + \frac{\sqrt{3}}{2}x_1y_3 + \\ \frac{\sqrt{3}}{2}x_3y_1 - \frac{\sqrt{3}}{2}x_3y_3 + \frac{3}{4}y_1^2 - \frac{3}{2}y_1y_3 + \frac{3}{4}y_3^2 + \frac{1}{4}x_2^2 - \frac{1}{2}x_2x_1 + \frac{1}{4}x_1^2 \\ \frac{\sqrt{3}}{2}x_2y_2 - \frac{\sqrt{3}}{2}x_2y_1 - \frac{\sqrt{3}}{2}x_1y_2 + \frac{\sqrt{3}}{2}x_1y_1 + \frac{3}{4}y_2^2 - \frac{3}{2}y_2y_1 + \frac{3}{4}y_1^2\end{array}\right\} \tag{4.47}$$

$$V_T = \frac{k}{2}\left\{\begin{array}{l}\frac{5}{4}x_3^2 - \frac{5}{4}x_2^2 + \frac{1}{2}x_1^2 + \frac{3}{2}y_1^2 + \frac{3}{4}y_2^2 + \frac{3}{4}y_3^2 - \frac{1}{2}x_2x_1 - \frac{1}{2}x_1x_3 - 2x_3x_2 + \\ \left(\frac{\sqrt{3}}{2}x_1y_1 - \frac{\sqrt{3}}{2}x_1y_1\right) - \frac{\sqrt{3}}{2}x_1y_2 + \frac{\sqrt{3}}{2}x_1y_3 - \frac{\sqrt{3}}{2}x_2y_1 + \frac{\sqrt{3}}{2}x_2y_2 + \\ \frac{\sqrt{3}}{2}x_3y_1 - \frac{\sqrt{3}}{2}x_3y_3 - \frac{3}{2}y_2y_1 - \frac{3}{2}y_1y_3\end{array}\right\} \tag{4.48}$$

Then the Lagrangian of the system is

$$L = T + V_T\left\{\begin{array}{l}\frac{m}{2}\left[\dot{x}_1^2 + \dot{y}_1^2 + \dot{x}_2^2 + \dot{y}_2^2 + \dot{x}_3^2 + \dot{y}_3^2\right] + \\ \frac{k}{2}\left[\begin{array}{l}\frac{5}{4}x_3^2 - \frac{5}{4}x_2^2 + \frac{1}{2}x_1^2 + \frac{3}{2}y_1^2 + \frac{3}{4}y_2^2 + \\ \frac{3}{4}y_3^2 - \frac{1}{2}x_2x_1 - \frac{1}{2}x_1x_3 - 2x_3x_2 - \\ \frac{\sqrt{3}}{2}x_1y_2 + \frac{\sqrt{3}}{2}x_1y_3 - \frac{\sqrt{3}}{2}x_2y_1 + \\ \frac{\sqrt{3}}{2}x_2y_2 + \frac{\sqrt{3}}{2}x_3y_1 - \frac{\sqrt{3}}{2}x_3y_3 - \\ \frac{3}{2}y_2y_1 - \frac{3}{2}y_1y_3\end{array}\right]\end{array}\right\} \tag{4.49}$$

Applying the Lagrange equation, which has the following form,

$$\frac{d}{dt}\left(\frac{\partial L}{\partial \dot{x}_k}\right) - \frac{\partial L}{\partial x_k} = 0 \quad j = 1, 2, 3 \ldots \tag{4.50}$$

Applying the above equation for the displacement coordinate x_1 of the mass m_1:

$$\frac{\partial L}{\partial \dot{x}_1} = m\dot{x}_1 \rightarrow \frac{d}{dt}\left(\frac{\partial L}{\partial \dot{x}_1}\right) = m\ddot{x}_1 \tag{4.51}$$

$$\frac{\partial L}{\partial x_1} = \frac{k}{2}\left[x_1 - \tfrac{1}{2}x_2 - \tfrac{1}{2}x_3 - \tfrac{\sqrt{3}}{2}y_2 + \tfrac{\sqrt{3}}{2}y_3\right] \tag{4.52}$$

Thus;

$$\frac{d}{dt}\left(\frac{\partial L}{\partial \dot{x}_1}\right) - \frac{\partial L}{\partial x_1} = 0 \tag{4.53}$$

$$m\ddot{x}_1 + \frac{k}{2}\left[x_1 - \tfrac{1}{2}x_2 - \tfrac{1}{2}x_3 - \tfrac{\sqrt{3}}{2}y_2 + \tfrac{\sqrt{3}}{2}y_3\right] = 0 \tag{4.54}$$

For the displacement coordinate y_1 of the mass m_1:

$$\begin{aligned}
&\tfrac{\partial L}{\partial \dot{y}_1} = m\dot{y}_1 \rightarrow \tfrac{d}{dt}\left(\tfrac{\partial L}{\partial \dot{y}_1}\right) = m\ddot{y}_1 \\
&\tfrac{\partial L}{\partial y_1} = \tfrac{k}{2}\left[3y_1 - \tfrac{\sqrt{3}}{2}x_2 + \tfrac{\sqrt{3}}{2}x_3 - \tfrac{3}{2}y_2 - \tfrac{3}{2}y_3\right] \\
&\tfrac{d}{dt}\left(\tfrac{\partial L}{\partial \dot{y}_1}\right) - \tfrac{\partial L}{\partial y_1} = 0 \\
&m\ddot{y}_1 + \tfrac{k}{2}\left[3y_1 - \tfrac{\sqrt{3}}{2}x_2 + \tfrac{\sqrt{3}}{2}x_3 - \tfrac{3}{2}y_2 - \tfrac{3}{2}y_3\right] F = 0
\end{aligned} \tag{4.55}$$

For the displacement coordinate x_2 of the mass m_2, it is obtained

$$\begin{aligned}
&\tfrac{\partial L}{\partial \dot{x}_2} = m\dot{x}_2 \rightarrow \tfrac{d}{dt}\left(\tfrac{\partial L}{\partial \dot{x}_2}\right) = m\ddot{x}_2 \\
&\tfrac{\partial L}{\partial x_2} = \tfrac{k}{2}\left[\tfrac{5}{2}x_2 - \tfrac{1}{2}x_1 - 2x_3 - \tfrac{\sqrt{3}}{2}y_1 + \tfrac{\sqrt{3}}{2}y_2\right] \\
&\tfrac{d}{dt}\left(\tfrac{\partial L}{\partial \dot{x}_2}\right) - \tfrac{\partial L}{\partial x_2} = 0 \\
&m\ddot{x}_2 + \tfrac{k}{2}\left[\tfrac{5}{2}x_2 - \tfrac{1}{2}x_1 - 2x_3 - \tfrac{\sqrt{3}}{2}y_1 + \tfrac{\sqrt{3}}{2}y_2\right] = 0
\end{aligned} \tag{4.56}$$

For the displacement coordinate y_2 of the mass m_2, it is obtained

$$\begin{aligned}
&\frac{\partial L}{\partial \dot{y}_2} = m\dot{y}_2 \rightarrow \frac{d}{dt}\left(\frac{\partial L}{\partial \dot{y}_2}\right) = m\ddot{y}_2 \\
&\frac{\partial L}{\partial y_2} = \frac{k}{2}\left[\frac{3}{2}y_2 - \frac{\sqrt{3}}{2}x_1 + \frac{\sqrt{3}}{2}x_2 - \frac{3}{2}y_1\right] \\
&\frac{d}{dt}\left(\frac{\partial L}{\partial \dot{y}_2}\right) - \frac{\partial L}{\partial y_2} = 0 \\
&m\ddot{y}_2 + \frac{k}{2}\left[\frac{3}{2}y_2 - \frac{\sqrt{3}}{2}x_1 + \frac{\sqrt{3}}{2}x_2 - \frac{3}{2}y_1\right] = 0
\end{aligned} \tag{4.57}$$

For the displacement coordinate x_3 of the mass m_3, it is obtained

$$\begin{aligned}
&\frac{\partial L}{\partial \dot{x}_3} = m\dot{x}_3 \rightarrow \frac{d}{dt}\left(\frac{\partial L}{\partial \dot{x}_3}\right) = m\ddot{x}_3 \\
&\frac{\partial L}{\partial x_3} = \frac{k}{2}\left[\frac{5}{2}x_3 - \frac{1}{2}x_1 - 2x_2 + \frac{\sqrt{3}}{2}y_1 - \frac{\sqrt{3}}{2}y_3\right] \\
&\frac{d}{dt}\left(\frac{\partial L}{\partial \dot{x}_3}\right) - \frac{\partial L}{\partial x_3} = 0 \\
&m\ddot{x}_3 + \frac{k}{2}\left[\frac{5}{2}x_3 - \frac{1}{2}x_1 - 2x_2 + \frac{\sqrt{3}}{2}y_1 - \frac{\sqrt{3}}{2}y_3\right] = 0
\end{aligned} \tag{4.58}$$

For the displacement coordinate y_3 of the mass m_3, it is obtained

$$\begin{aligned}
&\frac{\partial L}{\partial \dot{y}_3} = m\dot{y}_3 \rightarrow \frac{d}{dt}\left(\frac{\partial L}{\partial \dot{y}_3}\right) = m\ddot{y}_3 \\
&\frac{\partial L}{\partial y_3} = \frac{k}{2}\left[\frac{3}{2}y_3 + \frac{\sqrt{3}}{2}x_1 - \frac{\sqrt{3}}{2}x_3 - \frac{3}{2}y_1\right] \\
&\frac{d}{dt}\left(\frac{\partial L}{\partial \dot{y}_3}\right) - \frac{\partial L}{\partial y_3} = 0 \\
&m\ddot{y}_3 + \frac{k}{2}\left[\frac{3}{2}y_3 + \frac{\sqrt{3}}{2}x_1 - \frac{\sqrt{3}}{2}x_3 - \frac{3}{2}y_1\right] = 0
\end{aligned} \tag{4.59}$$

It recognizes that for differential equations with constant coefficients, the solutions involve the exponential function. Thus, It can propose the following expression as a solution:

$$x_j = X_j e^{i\omega t} \Rightarrow \dot{x}_j = i\omega X_j e^{i\omega t} \Rightarrow \ddot{x}_j = -\omega^2 X_j e^{i\omega t} \tag{4.60}$$

for $j = 1, 2, 3, \ldots$ Similarly for y, we have

$$y_j = Y_j e^{i\omega t} \Rightarrow \dot{y}_j = i\omega Y_j e^{i\omega t} \Rightarrow \ddot{y}_j = -\omega^2 Y_j e^{i\omega t} \tag{4.61}$$

Replacing Eqs. 4.60 and 4.61 into Eqs. 4.54, 4.56, 4.58, 4.55, 4.57, and 4.59, It is obtained

$$m\ddot{x}_1 + \frac{k}{2}\left[x_1 - \frac{1}{2}x_2 - \frac{1}{2}x_3 - \frac{\sqrt{3}}{2}y_2 + \frac{\sqrt{3}}{2}y_3\right] = 0 \tag{4.62}$$

$$-m\omega^2 X_1 e^{i\omega t} + \frac{k}{2}\left[X_1 e^{i\omega t} - \frac{1}{2}X_2 e^{i\omega t} - \frac{1}{2}X_3 e^{i\omega t} - \frac{\sqrt{3}}{2}Y_2 e^{i\omega t} + \frac{\sqrt{3}}{2}Y_3 e^{i\omega t}\right] = 0 \tag{4.63}$$

$$m\ddot{y}_1 + \frac{k}{2}\left[3y_1 - \frac{\sqrt{3}}{2}x_2 + \frac{\sqrt{3}}{2}x_3 - \frac{3}{2}y_2 - \frac{3}{2}y_3\right] = 0 \tag{4.64}$$

$$-m\omega^2 Y_1 e^{i\omega t} + \frac{k}{2}\left[3Y_1 e^{i\omega t} - \frac{\sqrt{3}}{2}X_2 e^{i\omega t} + \frac{\sqrt{3}}{2}X_3 e^{i\omega t} - \frac{3}{2}Y_2 e^{i\omega t} - \frac{3}{2}Y_3 e^{i\omega t}\right] = 0 \tag{4.65}$$

$$m\ddot{x}_2 + \frac{k}{2}\left[\frac{5}{2}x_2 - \frac{1}{2}x_1 - 2x_3 - \frac{\sqrt{3}}{2}y_1 + \frac{\sqrt{3}}{2}y_2\right] = 0 \tag{4.66}$$

$$-m\omega^2 X_2 e^{i\omega t} + \frac{k}{2}\left[\frac{5}{2}X_2 e^{i\omega t} - \frac{1}{2}X_1 e^{i\omega t} - 2X_3 e^{i\omega t} - \frac{\sqrt{3}}{2}Y_1 e^{i\omega t} + \frac{\sqrt{3}}{2}Y_2 e^{i\omega t}\right] = 0 \tag{4.67}$$

$$m\ddot{y}_2 + \frac{k}{2}\left[\frac{3}{2}y_2 - \frac{\sqrt{3}}{2}x_1 + \frac{\sqrt{3}}{2}x_2 - \frac{3}{2}y_1\right] = 0 \tag{4.68}$$

$$-m\omega^2 Y_2 e^{i\omega t} + \frac{k}{2}\left[\frac{3}{2}Y_2 e^{i\omega t} - \frac{\sqrt{3}}{2}X_1 e^{i\omega t} + \frac{\sqrt{3}}{2}X_2 e^{i\omega t} - \frac{3}{2}Y_1 e^{i\omega t}\right] = 0 \tag{4.69}$$

$$m\ddot{x}_3 + \frac{k}{2}\left[\frac{5}{2}x_3 - \frac{1}{2}x_1 - 2x_2 + \frac{\sqrt{3}}{2}y_1 - \frac{\sqrt{3}}{2}y_3\right] = 0 \tag{4.70}$$

$$-m\omega^2 X_3 e^{i\omega t} + \frac{k}{2}\left[\frac{5}{2}X_3 e^{i\omega t} - \frac{1}{2}X_1 e^{i\omega t} - 2X_2 e^{i\omega t} + \frac{\sqrt{3}}{2}Y_1 e^{i\omega t} - \frac{\sqrt{3}}{2}Y_3 e^{i\omega t}\right] = 0 \tag{4.71}$$

$$m\ddot{y}_3 + \frac{k}{2}\left[\frac{3}{2}y_3 + \frac{\sqrt{3}}{2}x_1 - \frac{\sqrt{3}}{2}x_3 - \frac{3}{2}y_1\right] = 0 \tag{4.72}$$

$$-m\omega^2 Y_3 e^{i\omega t} + \frac{k}{2}\left[\frac{3}{2}Y_3 e^{i\omega t} + \frac{\sqrt{3}}{2}X_1 e^{i\omega t} - \frac{\sqrt{3}}{2}X_3 e^{i\omega t} - \frac{3}{2}Y_1 e^{i\omega t}\right] = 0 \tag{4.73}$$

Replacing in these expressions $\lambda = \frac{m\omega^2}{k}$, we obtain the following equations:

$$-k\lambda X_1 e^{i\omega t} + \frac{k}{2}e^{i\omega t}\left[X_1 - \frac{1}{2}X_2 - \frac{1}{2}X_3 - \frac{\sqrt{3}}{2}Y_2 + \frac{\sqrt{3}}{2}Y_3\right] = 0 \tag{4.74}$$

$$-\lambda X_1 + \frac{1}{2}\left[X_1 - \frac{1}{2}X_2 - \frac{1}{2}X_3 - \frac{\sqrt{3}}{2}Y_2 + \frac{\sqrt{3}}{2}Y_3\right] = 0 \tag{4.75}$$

$$X_1\left(\frac{1}{2} - \lambda\right) - \frac{1}{4}X_2 - \frac{\sqrt{3}}{4}Y_2 - \frac{1}{4}X_3 + \frac{\sqrt{3}}{4}Y_3 = 0 \tag{4.76}$$

$$-k\lambda Y_1 e^{i\omega t} + \frac{k}{2}e^{i\omega t}\left[3Y_1 - \frac{\sqrt{3}}{2}X_2 + \frac{\sqrt{3}}{2}X_3 - \frac{3}{2}Y_2 - \frac{3}{2}Y_3\right] = 0 \tag{4.77}$$

$$-\lambda Y_1 + \frac{1}{2}\left[3Y_1 - \frac{\sqrt{3}}{2}X_2 + \frac{\sqrt{3}}{2}X_3 - \frac{3}{2}Y_2 - \frac{3}{2}Y_3\right] = 0 \tag{4.78}$$

$$Y_1\left(\frac{3}{2} - \lambda\right) - \frac{\sqrt{3}}{4}X_2 - \frac{3}{4}Y_2 + \frac{\sqrt{3}}{4}X_3 - \frac{3}{4}Y_3 = 0 \tag{4.79}$$

$$-k\lambda X_2 e^{i\omega t} + \frac{k}{2}e^{i\omega t}\left[\frac{5}{2}X_2 - \frac{1}{2}X_1 - 2X_3 - \frac{\sqrt{3}}{2}Y_1 + \frac{\sqrt{3}}{2}Y_2\right] = 0 \tag{4.80}$$

$$-\lambda X_2 + \frac{1}{2}\left[\frac{5}{2}X_2 - \frac{1}{2}X_1 - 2X_3 - \frac{\sqrt{3}}{2}Y_1 + \frac{\sqrt{3}}{2}Y_2\right] = 0 \tag{4.81}$$

$$\frac{1}{4}X_1 - \frac{\sqrt{3}}{4}Y - X_2\left(\frac{5}{4} - \lambda\right) + \frac{\sqrt{3}}{4}Y_2 - X_3 = 0 \tag{4.82}$$

$$-k\lambda Y_2 e^{i\omega t} + \frac{k}{2}e^{i\omega t}\left[\frac{3}{2}Y_2 - \frac{\sqrt{3}}{2}X_1 + \frac{\sqrt{3}}{2}X_2 - \frac{3}{2}Y_1\right] = 0 \tag{4.83}$$

$$-\lambda Y_2 + \frac{1}{2}\left[\frac{3}{2}Y_2 - \frac{\sqrt{3}}{2}X_1 + \frac{\sqrt{3}}{2}X_2 - \frac{3}{2}Y_1\right] = 0 \tag{4.84}$$

$$-\frac{\sqrt{3}}{4}X_1 - \frac{3}{4}Y_1 + \frac{\sqrt{3}}{4}X_2 + Y_2\left(\frac{3}{4} - \lambda\right) = 0 \tag{4.85}$$

$$-k\lambda X_3 e^{i\omega t} + \frac{k}{2}e^{i\omega t}\left[\frac{5}{2}X_3 - \frac{1}{2}X_1 - 2X_2 + \frac{\sqrt{3}}{2}Y_1 - \frac{\sqrt{3}}{2}Y_3\right] = 0 \tag{4.86}$$

$$-\lambda X_3 + \frac{1}{2}\left[\frac{5}{2}X_3 - \frac{1}{2}X_1 - 2X_2 + \frac{\sqrt{3}}{2}Y_1 - \frac{\sqrt{3}}{2}Y_3\right] = 0 \tag{4.87}$$

$$-\frac{1}{4}X_1 + \frac{\sqrt{3}}{4}Y_1 - X_2 + X_3\left(\frac{5}{4} - \lambda\right) - \frac{\sqrt{3}}{4}Y_3 = 0 \tag{4.88}$$

$$-k\lambda Y_3 e^{i\omega t} + \frac{k}{2}e^{i\omega t}\left[\frac{3}{2}Y_3 + \frac{\sqrt{3}}{2}X_1 - \frac{\sqrt{3}}{2}X_3 - \frac{3}{2}Y_1\right] = 0 \tag{4.89}$$

$$-\lambda Y_3 + \frac{1}{2}\left[\frac{3}{2}Y_3 + \frac{\sqrt{3}}{2}X_1 - \frac{\sqrt{3}}{2}X_3 - \frac{3}{2}Y_1\right] = 0 \tag{4.90}$$

$$\frac{\sqrt{3}}{4}X_1 - \frac{3}{4}Y_1 - \frac{\sqrt{3}}{4}X_3 + Y_3\left(\frac{3}{4} - \lambda\right) = 0 \tag{4.91}$$

Thus, this system of equations leads to the following secular equation:

$$\begin{vmatrix} \left(\frac{1}{2}-\lambda\right) & 0 & -\frac{1}{4} & -\frac{\sqrt{3}}{4} & -\frac{1}{4} & \frac{\sqrt{3}}{4} \\ 0 & \left(\frac{3}{2}-\lambda\right) & -\frac{\sqrt{3}}{4} & -\frac{3}{4} & \frac{\sqrt{3}}{4} & -\frac{3}{4} \\ -\frac{1}{4} & -\frac{\sqrt{3}}{4} & \left(\frac{5}{4}-\lambda\right) & \frac{\sqrt{3}}{4} & -1 & 0 \\ -\frac{\sqrt{3}}{4} & -\frac{3}{4} & \frac{\sqrt{3}}{4} & \left(\frac{3}{4}-\lambda\right) & 0 & 0 \\ -\frac{1}{4} & \frac{\sqrt{3}}{4} & -1 & 0 & \left(\frac{5}{4}-\lambda\right) & -\frac{\sqrt{3}}{4} \\ \frac{\sqrt{3}}{4} & -\frac{3}{4} & 0 & 0 & -\frac{\sqrt{3}}{4} & \left(\frac{3}{4}-\lambda\right) \end{vmatrix} = \tag{4.92}$$

The characteristic polynomial, after performing the appropriate algebra, is

$$\lambda^6 - 6\lambda^5 + \frac{45}{4}\lambda^4 - \frac{27}{4}\lambda^3 = 0 \tag{4.93}$$

Plotting this polynomial, we can obtain its roots: The roots are the points where the curve intersects the λ axis, so it is observed that the roots are: $\lambda = 3$, $\lambda = \frac{3}{2}$, $\lambda = 0$. Since the polynomial is of the sixth order and only three roots are found, it is clear that some of them are multiple roots. Therefore, the roots are (Fig. 4.7):

$$\lambda = 0, 0, 0, \frac{3}{2}, \frac{3}{2}, 3 \tag{4.94}$$

These values of λ, when replaced in the expression $\lambda = \frac{m\omega^2}{k}$, give the following frequency values:

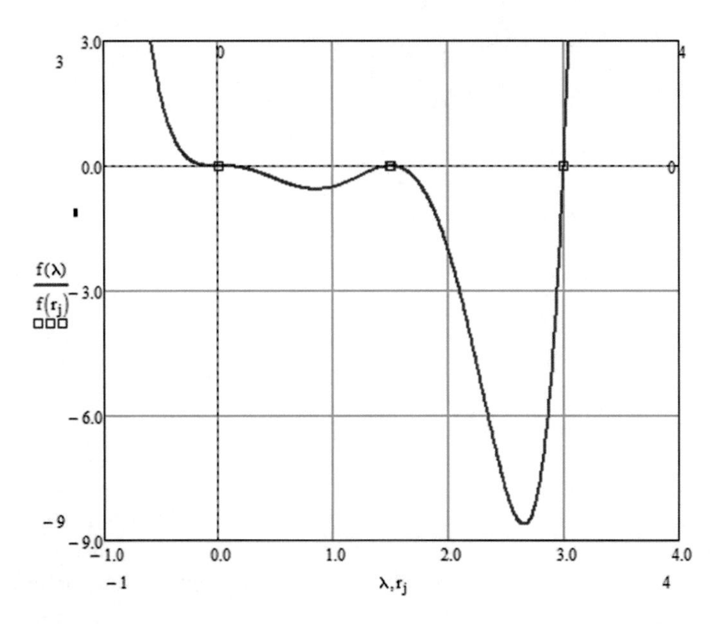

Fig. 4.7 Roots of the polynomial

$$0 = \frac{m\omega^2}{k} \Rightarrow \omega^2 = 0 \Rightarrow \upsilon_2 = 0 \tag{4.95}$$

$$\frac{3}{2} = \frac{m\omega^2}{k} \Rightarrow \omega_1 = \sqrt{\frac{3}{2}\frac{k}{m}} \Rightarrow \upsilon_1 = \frac{1}{2\pi}\sqrt{\frac{3}{2}\frac{k}{m}} \tag{4.96}$$

$$3 = \frac{m\omega^2}{k} \Rightarrow \omega_3 = \sqrt{\frac{3k}{m}} \Rightarrow \upsilon_3 = \frac{1}{2\pi}\sqrt{\frac{3k}{m}} \tag{4.97}$$

Finally, by substituting the values λ_j from Eq. 4.94 into Eqs. 4.76, 4.79, 4.82, 4.85, 4.88, and 4.94, we can determine the amplitudes X_1, X_2, X_3 and Y_1, Y_2, Y_3. Since the λ_j are the eigenvalues, they are substituted into the linear equations where these equations are solved for the X_i and Y_i.

To better manage the equations, we take the coordinates X_i and Y_i as $X_1 = a_1, Y_1 = a_2, X_2 = a_3, Y_2 = a_4, X_3 = a_5, Y_3 = a_6$, so that the equations can be organized as follows:

$$a_1\left(\frac{1}{2} - \lambda\right) + 0a_2 - \frac{1}{4}a_3 - \frac{\sqrt{3}}{4}a_4 - \frac{1}{4}a_5 + \frac{\sqrt{3}}{4}a_6 = 0 \tag{4.98}$$

$$0a_1 + a_2\left(\frac{3}{2} - \lambda\right) - \frac{\sqrt{3}}{4}a_3 - \frac{3}{4}a_4 + \frac{\sqrt{3}}{4}a_5 - \frac{3}{4}a_6 = 0 \tag{4.99}$$

$$-\frac{1}{4}a_1 - \frac{\sqrt{3}}{4}a_2 + a_3\left(\frac{5}{4} - \lambda\right) + \frac{\sqrt{3}}{4}a_4 - a_5 + 0a_6 = 0 \tag{4.100}$$

$$-\frac{\sqrt{3}}{4}a_1 - \frac{3}{4}a_2 + \frac{\sqrt{3}}{4}a_3 + a_4\left(\frac{3}{4} - \lambda\right) + 0a_5 + 0a_6 = 0 \tag{4.101}$$

$$-\frac{1}{4}a_1 + \frac{\sqrt{3}}{4}a_2 - a_3 + 0a_4 + a_5\left(\frac{5}{4} - \lambda\right) - \frac{\sqrt{3}}{4}a_6 = 0 \tag{4.102}$$

$$\frac{\sqrt{3}}{4}a_1 - \frac{3}{4}a_2 + 0a_3 + 0a_4 - \frac{\sqrt{3}}{4}a_5 + a_6\left(\frac{3}{4} - \lambda\right) = 0 \tag{4.103}$$

These are six equations with six unknowns, but we can take an alternative method. For example, since a_1 can take the value of zero or a finite value, in this case, this finite value can be one. This would ensure the normalization of the displacements with respect to a_1, and we would solve the equations for the remaining values a_2, a_3, a_4, a_5 and a_6. Thus, Eqs. 4.98 become

(a) For λ = 0:

$$0a_2 - \frac{1}{4}a_3 - \frac{\sqrt{3}}{4}a_4 - \frac{1}{4}a_5 + \frac{\sqrt{3}}{4}a_6 = \frac{1}{2} \tag{4.104}$$

$$a_2\left(\frac{3}{2}\right) - \frac{\sqrt{3}}{4}a_3 - \frac{3}{4}a_4 + \frac{\sqrt{3}}{4}a_5 - \frac{3}{4}a_6 = 0 \tag{4.105}$$

$$-\frac{\sqrt{3}}{4}a_2 + \frac{5}{4}a_3 + \frac{\sqrt{3}}{4}a_4 - a_5 + 0a_6 = \frac{1}{4} \tag{4.106}$$

$$-\frac{3}{4}a_2 + \frac{\sqrt{3}}{4}a_3 + \frac{3}{4}a_4 + 0a_5 + 0a_6 = \frac{\sqrt{3}}{4} \tag{4.107}$$

$$\frac{\sqrt{3}}{4}a_2 - a_3 + 0a_4 + \frac{5}{4}a_5 - \frac{\sqrt{3}}{4}a_6 = \frac{1}{4} \tag{4.108}$$

$$-\frac{3}{4}a_2 + 0a_3 + 0a_4 - \frac{\sqrt{3}}{4}a_5 + \frac{3}{4}a_6 = -\frac{\sqrt{3}}{4} \tag{4.109}$$

Solving the system, we find that

$$a_2 = 0 \tag{4.110}$$

and substituting this value into Eqs. 4.101 and 4.103, we obtain

$$a_4 = \frac{1 - a_3}{\sqrt{3}} \tag{4.111}$$

$$a_6 = \frac{a_5 - 1}{\sqrt{3}} \tag{4.112}$$

Replacing these values obtained in Eq. 4.99, it is obtained

$$a_3 = a_5 \tag{4.113}$$

Since it is advisable to use unit displacements for each atom within the molecule separately, it is obtained

$$\sqrt{a_1^2 + a_2^2 = 1} \tag{4.114}$$

$$\sqrt{a_3^2 + a_4^2 = 1} \tag{4.115}$$

$$\sqrt{a_5^2 + a_6^2 = 1} \tag{4.116}$$

We obtain

$$\sqrt{a_3^2 + a_4^2} = \sqrt{a_3^2 + \left(\frac{1 - a_3}{\sqrt{3}}\right)^2} = \sqrt{\frac{4}{3}a_3^2 - \frac{2}{3}a_3 + \frac{1}{3} = 1} \tag{4.117}$$

Then:

$$4a_3^2 - 2a_3 - 2 = 0 \tag{4.118}$$

The solution of this polynomial indicates that for a_3, there are two solutions, which are $a_3 = 1$ and $a_3 = \frac{1}{2}$. These results indicate two groups of values for a_1, a_2, a_3, a_4, a_5 and a_6 obtained from Eqs. 4.110 to 4.116:

$$a_1 = a_3 = a_5 = 1, a_2 = a_4 = a_6 = 0 \tag{4.119}$$

and

$$a_1 = 1, a_2 = 0, a_3 = -\frac{1}{2}, a_4 = \frac{\sqrt{3}}{2}, a_5 = -\frac{1}{2}, a_6 = -\frac{\sqrt{3}}{2} \tag{4.120}$$

Normalizing these groups of values with

$$|\hat{n}| = \left|\frac{n}{|n|}\right| = \frac{\mathbf{a}}{\sqrt{\sum_{i=1}^{6} a_i^2}} \tag{4.121}$$

Then the normalization factor for Eq. 4.119 is

$$\sqrt{\sum_{i=1}^{6} a_i^2} = \sqrt{1^2 + 1^2 + 1^2} = \sqrt{3} \tag{4.122}$$

And for Eq. 4.120 it is

$$\sqrt{\sum_{i=1}^{6} a_i^2} = \sqrt{1^2 + 2\left(\frac{1}{2}\right)^2 + 2\left(\frac{\sqrt{3}}{2}\right)^2} = \sqrt{3} \tag{4.123}$$

Therefore, the normalized displacement values from (4.119) are

$$a_1 = \frac{1}{\sqrt{3}}, a_2 = 0, a_3 = \frac{1}{\sqrt{3}}, a_4 = 0, a_5 = \frac{1}{\sqrt{3}}, a_6 = 0 \tag{4.124}$$

and the normalized displacements from (4.120) are

$$a_1 = \frac{1}{\sqrt{3}}, a_2 = 0, a_3 = -\frac{1}{2\sqrt{3}}, a_4 = \frac{1}{2}, a_5 = -\frac{1}{2\sqrt{3}}, a_6 = -\frac{1}{2} \tag{4.125}$$

When taking the equation of unit displacements $\sqrt{a_1^2 + a_2^2} = 1$ and assigning $a_2 = 1$, then $a_1 = 0$, therefore we obtain the already normalized displacement values (Table 4.1):

$$a_1 = 0, a_2 = \frac{1}{\sqrt{3}}, a_3 = 0, a_4 = \frac{1}{\sqrt{3}}, a_5 = 0, a_6 = \frac{1}{\sqrt{3}} \tag{4.126}$$

The figure representing such atomic movements is (Fig. 4.8):

(b) For $\lambda = \frac{3}{2}$:

$$0a_2 - \frac{1}{4}a_3 - \frac{\sqrt{3}}{4}a_4 - \frac{1}{4}a_5 + \frac{\sqrt{3}}{4}a_6 = 1 \tag{4.127}$$

Table 4.1 Displacement of the atoms for $\lambda = 0$

Position of the atoms for $\lambda = 0$		
Átom 1	Átom 2	Átom 3
$a_1 = \frac{1}{\sqrt{3}}, a_2 = 0$	$a_3 = \frac{1}{\sqrt{3}}, a_4 = 0$	$a_5 = \frac{1}{\sqrt{3}}, a_6 = 0$
$a_1 = \frac{1}{\sqrt{3}}, a_2 = 0$	$a_3 = -\frac{1}{2\sqrt{3}}, a_4 = \frac{1}{2}$	$a_5 = -\frac{1}{2\sqrt{3}}, a_6 = -\frac{1}{2}$
$a_1 = 0, a_2 = \frac{1}{\sqrt{3}}$	$a_3 = 0, a_4 = \frac{1}{\sqrt{3}}$	$a_5 = 0, a_6 = \frac{1}{\sqrt{3}}$

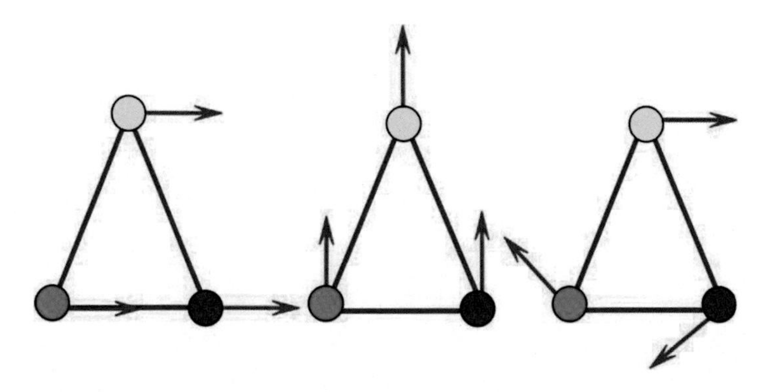

Fig. 4.8 Normal modes for Planar Equilateral Triangle

Table 4.2 Displacement of the atoms for $\lambda = \frac{3}{2}$

Position of atoms for $\lambda = \frac{3}{2}$		
Atom 1	Atom 2	Atom 3
$a_1 = \frac{1}{\sqrt{3}}, a_2 = 0$	$a_3 = -\frac{1}{2\sqrt{3}}, a_4 = -\frac{1}{2}$	$a_5 = -\frac{1}{2\sqrt{3}}, a_6 = \frac{1}{2}$
$a_1 = 0, a_2 = \frac{1}{\sqrt{3}}$	$a_3 = \frac{1}{2}, a_4 = -\frac{1}{2\sqrt{3}} =$	$a_5 = -\frac{1}{2}, a_6 = -\frac{1}{2\sqrt{3}}$

$$0a_2 - \frac{\sqrt{3}}{4}a_3 - \frac{3}{4}a_4 + \frac{\sqrt{3}}{4}a_5 - \frac{3}{4}a_6 = 0 \tag{4.128}$$

$$-\frac{\sqrt{3}}{4}a_2 - \frac{1}{4}a_3 + \frac{\sqrt{3}}{4}a_4 - a_5 + 0a_6 = \frac{1}{4} \tag{4.129}$$

$$-\frac{3}{4}a_2 + \frac{\sqrt{3}}{4}a_3 - \frac{3}{4}a_4 + 0a_5 + 0a_6 = \frac{\sqrt{3}}{4} \tag{4.130}$$

$$\frac{\sqrt{3}}{4}a_2 - a_3 + 0a_4 - \frac{1}{4}a_5 - \frac{\sqrt{3}}{4}a_6 = \frac{1}{4} \tag{4.131}$$

$$-\frac{3}{4}a_2 + 0a_3 + 0a_4 - \frac{\sqrt{3}}{4}a_5 - \frac{3}{4}a_6 = -\frac{\sqrt{3}}{4} \tag{4.132}$$

The solution to this system of equations indicates that the displacements of the atoms follow the table below (Table 4.2).

(c) For $\lambda = 3$:

$$0a_2 - \frac{1}{4}a_3 - \frac{\sqrt{3}}{4}a_4 - \frac{1}{4}a_5 + \frac{\sqrt{3}}{4}a_6 = \frac{5}{2} \tag{4.133}$$

$$-\frac{3}{2}a_2 - \frac{\sqrt{3}}{4}a_3 - \frac{3}{4}a_4 + \frac{\sqrt{3}}{4}a_5 - \frac{3}{4}a_6 = 0 \tag{4.134}$$

$$-\frac{\sqrt{3}}{4}a_2 - \frac{7}{4}a_3 + \frac{\sqrt{3}}{4}a_4 - a_5 + 0a_6 = \frac{1}{4} \tag{4.135}$$

$$-\frac{3}{4}a_2 + \frac{\sqrt{3}}{4}a_3 - \frac{9}{4}a_4 + 0a_5 + 0a_6 = \frac{\sqrt{3}}{4} \tag{4.136}$$

$$\frac{\sqrt{3}}{4}a_2 - a_3 + 0a_4 - \frac{7}{4}a_5 - \frac{\sqrt{3}}{4}a_6 = \frac{1}{4} \tag{4.137}$$

$$-\frac{3}{4}a_2 + 0a_3 + 0a_4 - \frac{\sqrt{3}}{4}a_5 - \frac{9}{4}a_6 = -\frac{\sqrt{3}}{4} \tag{4.138}$$

Table 4.3 Displacement of the atoms for $\lambda = 3$

Position of the atoms for $\lambda = 3$		
Atom 1	Atom 2	Atom 3
$a_1 = 0, a_2 = \frac{1}{\sqrt{3}}$	$a_3 = -\frac{1}{2}, a_4 = -\frac{1}{2\sqrt{3}} =$	$a_5 = \frac{1}{2}, a_6 = -\frac{1}{2\sqrt{3}}$

The solution to this system of equations indicates that there is a unique solution, and the atomic displacements follow the coordinates specified in the following table (Table 4.3):

4.4 Matrix Method

The use of matrix methods in most cases simplifies the determination of the energies of the normal modes of vibration. When dealing with systems of coupled springs, the system of equations obtained involves coupling parameters, which with the help of the so-called normal coordinates q_i achieve decoupling and allow the determination of the vibrational energies associated with these coordinates.

To obtain the normal coordinates, we proceed to use the following transformation equation:

$$q_i = \sqrt{m} \sum_{j=1}^{3} a_{ij} x_j \tag{4.139}$$

where a_{ij} are the normalized amplitudes associated with the x coordinates and the vibrational frequencies ω_i. By using these types of coordinates in the potential energy V, the cross terms, such as $x_i x_j \rightarrow i \neq j$, are eliminated, and in such a case, the potential $V$and kinetic T energies are written as

$$V = \frac{1}{2} \sum_i \omega_i^2 q_i \quad T = \frac{1}{2} \sum_i \dot{q}_i^2 \tag{4.140}$$

As can be seen, the energy terms acquire a simpler form from a functional point of view.

4.4.1 Applications

(a) Linearly Coupled Three-Mass System

Let's consider the previously solved problem of three equal masses m coupled by two equal springs with a spring constant k, as illustrated in Fig. 4.18.

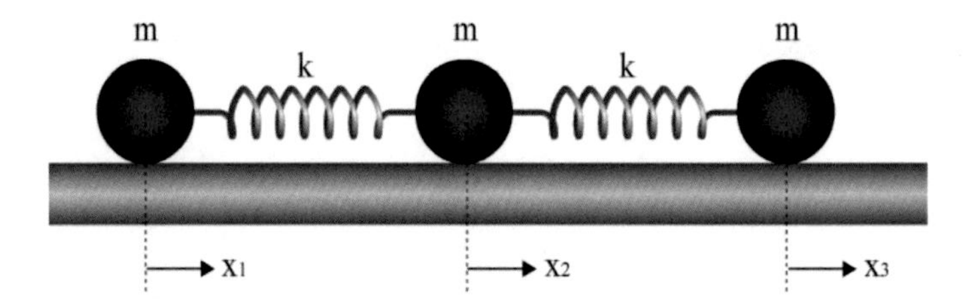

Fig. 4.9 3-mass coupled system

The differential equations of the system were obtained in the section on normal coordinates and are (Fig. 4.9):

$$\begin{aligned} m\ddot{x}_1 - k(x_2 - x_1) &= 0 \quad (a) \\ m\ddot{x}_2 + k(x_2 - x_1) - k(x_3 - x_2) &= 0 \quad (b) \\ m\ddot{x}_3 + k(x_3 - x_2) &= 0 \quad (c) \end{aligned} \tag{4.141}$$

Let's apply the matrix method. In this case, we define the normal coordinates by

$$q_s = \sqrt{m}\sum_{i=j}^{3} a_s x_r \Rightarrow a_i \rightarrow a_{ij} \tag{4.142}$$

$$q_1 = \sqrt{m}(a_1x_1 + a_2x_2 + a_3x_3) \Rightarrow \lambda_1 \tag{4.143}$$

$$q_1 = \sqrt{\frac{m}{3}}(x_1 + x_2 + x_3) \tag{4.144}$$

$$q_2 = \sqrt{m}(a_1x_1 + a_2x_2 + a_3x_3) \Rightarrow \lambda_2 = 1 \tag{4.145}$$

$$q_2 = \sqrt{m}\left(\frac{1}{\sqrt{2}}x_1 - \frac{1}{\sqrt{2}}x_3\right) \tag{4.146}$$

$$q_2 = \sqrt{\frac{m}{2}}(x_1 - x_3) \tag{4.147}$$

$$q_3 = \sqrt{m}\left(\frac{1}{\sqrt{6}}x_1 - \frac{2}{\sqrt{6}}x_2 + \frac{1}{\sqrt{6}}x_3\right) \Rightarrow \lambda_3 = 3 \tag{4.148}$$

$$q_3 = \sqrt{\frac{m}{6}}(x_1 - 2x_2 + x_3) \tag{4.149}$$

Adding the three differential equations (a), (b), and (c) written in (4.141), we obtain

$$m\ddot{x}_1 - k(x_2 - x_1) + m\ddot{x}_2 + k(x_2 - x_1) - k(x_3 - x_2) + m\ddot{x}_3 + k(x_3 - x_2) = 0$$

$$m\ddot{x}_1 + m\ddot{x}_2 + +m\ddot{x}_3 = 0 \rightarrow m\overset{\bullet\bullet}{\overline{(x_1 + x_2 + x_3)}} = 0 \tag{4.150}$$

And if we replace the normal coordinate q_1 from Eq. 4.144, we obtain

$$m\overset{\bullet\bullet}{\overline{(x_1 + x_2 + x_3)}} = m\overset{\bullet\bullet}{\overline{\left(\sqrt{\frac{3}{m}}q_1\right)}} = \sqrt{3m}\ddot{q}_1 = 0 \tag{4.151}$$

Since we are analyzing harmonic oscillatory motion, we can consider the general solution of the undamped motion for the coordinate q_n as

$$q_n = q_{0n}e^{i\omega_n t} \tag{4.152}$$

where the first and second derivatives with respect to time t are respectively:

$$\dot{q}_n = i\omega_n q_{0n}e^{i\omega_n t} = i\omega_n q_n \tag{4.153}$$

$$\ddot{q}_n = i\omega_n\dot{q}_n = i\omega_n i\omega_n q_n = i^2\omega_n^2 q_n = -\omega_n^2 q_n \tag{4.154}$$

In this case, for the coordinate q_1:

$$q_1 = q_{01}e^{i\omega_1 t} \;\; \text{y} \;\; \ddot{q}_1 = -\omega_1^2 q_1 \tag{4.155}$$

And comparing with Eq. 4.151, we obtain

$$\sqrt{3m}\ddot{q}_1 = 0 \rightarrow -\sqrt{3m}\omega_1^2 q_1 = 0 \rightarrow \text{si},\, \omega_1 = 0 \tag{4.156}$$

The result $\omega_1 = 0$ indicates no oscillatory motion, so this is seen as a complete translation of the three masses m, see Fig. 4.10:

As we see, one of the objectives of normal coordinates is to decouple the differential equations, each with an independent solution, which implies the natural frequencies or normal frequencies.

For the calculation of q_2, we use the Eq. (4.141), subtracting (a) from (c); that is

$$m\ddot{x}_1 - k(x_2 - x_1) - m\ddot{x}_3 - k(x_3 - x_2) = 0 \tag{4.157}$$

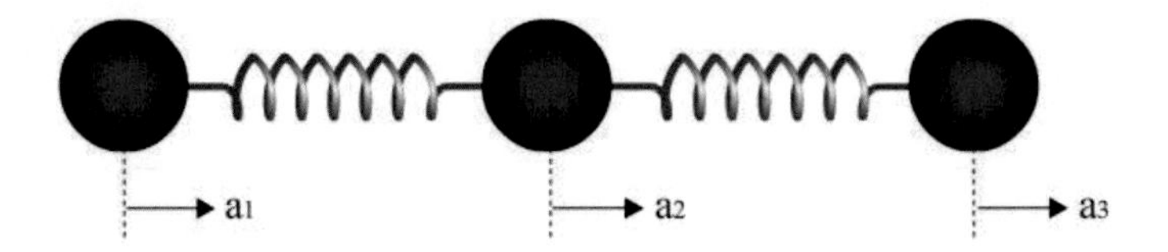

Fig. 4.10 Vibration of $\lambda_1 = 0$ (Traslación)

$$m\left(\ddot{x}_1 - \ddot{x}_3\right) - k\left(x_2 - x_1\right) - k\left(x_3 - x_2\right) = 0 \tag{4.158}$$

$$m\,\overset{\bullet\bullet}{\overline{\left(x_1 - x_3\right)}} + k\left(x_1 - x_3\right) = 0 \tag{4.159}$$

Using Eq. 4.147 and solving:

$$\left(x_1 - x_3\right) = \sqrt{\frac{2}{m}}q_2 \tag{4.160}$$

For $\lambda_2 = 1$

$$m\sqrt{\frac{2}{m}}\ddot{q}_2 + k\sqrt{\frac{2}{m}}q_2 = 0 \tag{4.161}$$

$$\ddot{q}_2 + \frac{k}{m}q_2 = 0 \tag{4.162}$$

Using the solution according to Eq. 4.152:

$$q_2 = q_{02}e^{i\omega_2 t} \tag{4.163}$$

Therefore:

$$\ddot{q}_2 = -\omega_2^2 q_2 \Rightarrow \; -\omega_2^2 q_2 \frac{k}{m} q_2 = 0 \Rightarrow \omega_2 = \sqrt{\frac{k}{m}} \tag{4.164}$$

For the system's motion with the normal coordinate q_2, we have a frequency $\omega_2 = \sqrt{\frac{k}{m}}$, and its motion is illustrated in Fig. 4.11.

For the calculation of coordinate q_3 in (4.141), where we have performed the operations: (a) -2(b)+(c):

$$m\ddot{x}_1 - k\left(x_2 - x_1\right) - 2\,m\ddot{x}_2 - 2k\left(x_2 - x_1\right) + 2k\left(x_3 - x_2\right) + m\ddot{x}_3 + k\left(x_3 - x_2\right) = 0 \tag{4.165}$$

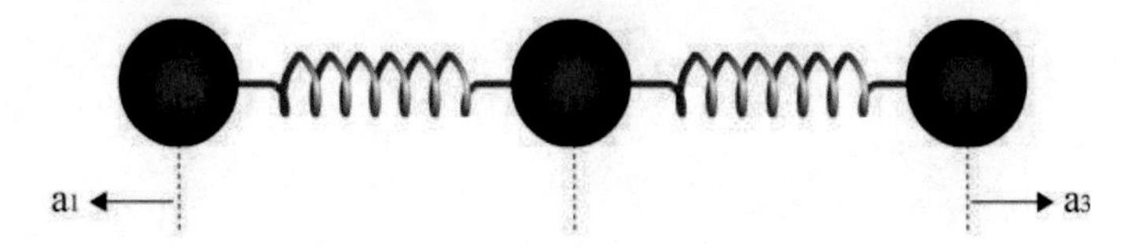

Fig. 4.11 Vibration of $\lambda_2 = 1$

$$m\,\overset{\bullet\bullet}{\overline{(x_1 - 2x_2 + x_3)}} - k\,(x_2 - x_1) - 2k\,(x_2 - x_1) + 2k\,(x_3 - x_2) + k\,(x_3 - x_2) = 0 \tag{4.166}$$

$$m\,\overset{\bullet\bullet}{\overline{(x_1 - 2x_2 + x_3)}} + 3k(x_1 - 2x_2 + x_3) = 0 \tag{4.167}$$

By using Eq. 4.149 for the coordinate q_3:

$$\sqrt{\frac{6}{m}}q_3 = (x_1 - 2x_2 + x_3) \tag{4.168}$$

And substituting into Eq. 4.167, it is obtained:

$$\ddot{q}_3 + \frac{3k}{m}q_3 = 0 \tag{4.169}$$

Let the solution according to Eq. 4.152 be

$$q_3 = q_{03}e^{i\omega_3 t} \tag{4.170}$$

and knowing that differentiating twice according to (4.154) gives

$$\ddot{q}_3 = -\omega_3^2 q_3 \tag{4.171}$$

therefore, substituting into Eq. 4.170 it is obtained:

$$-\omega_3^2 q_3 + \frac{3k}{m}q_3 = 0 \tag{4.172}$$

$$\omega_3 = \sqrt{\frac{3k}{m}} \tag{4.173}$$

This indicates that the third normal mode of vibration of the system with normal coordinate q_3 and frequency $\omega_3 = \sqrt{\frac{3k}{m}}$ oscillates according to Fig. 4.12:

Some conclusions that can be drawn from analyzing the results according to the obtained oscillation frequencies for each of the modes are

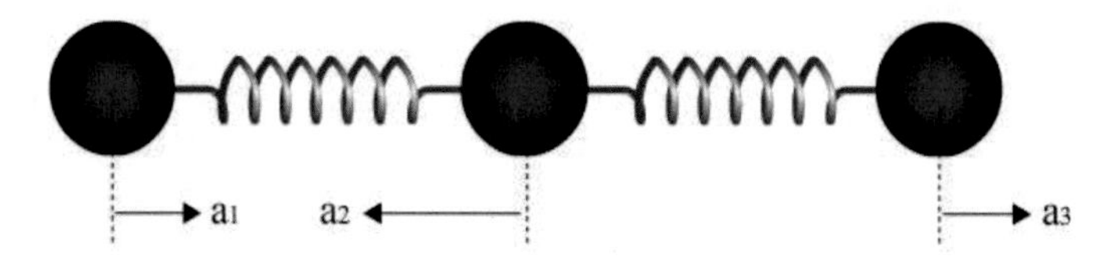

Fig. 4.12 Vibration of $\lambda_3 = 3$

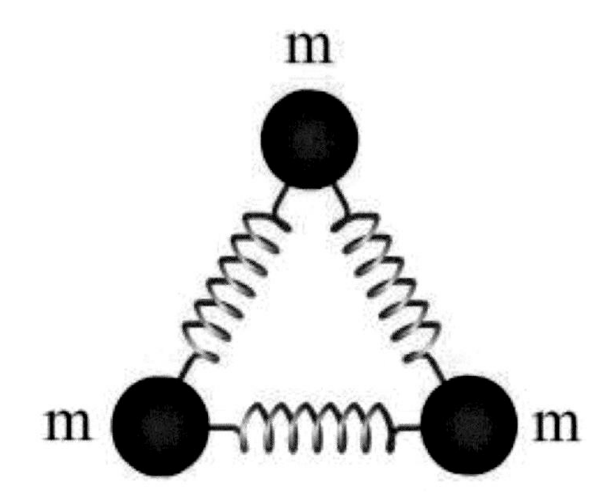

Fig. 4.13 System of 3 coupled masses in a planar equilateral triangle

- The antisymmetric normal modes have a higher frequency than the symmetric ones.
- The normal modes with a high degree of symmetry will have the lowest frequency.

(b) **System of three coupled masses forming an equilateral triangle**

This is one of the widely applied systems because the type of equilateral triangle structure is found widely in molecular systems, see Fig. 4.13.

Let the coordinate transformation be

$$q_s = \sqrt{m} \sum_{i=1}^{6} a_{si} x_i \tag{4.174}$$

Explicitly writing the equation, it is obtained

$$q_1 = \sqrt{m}(a_{11}x_1 + a_{12}x_2 + a_{13}x_3 + a_{14}x_4 + a_{15}x_5 + a_{16}x_6) \tag{4.175}$$

$$q_2 = \sqrt{m}(a_{21}x_1 + a_{22}x_2 + a_{23}x_3 + a_{24}x_4 + a_{25}x_5 + a_{26}x_6) \tag{4.176}$$

$$q_3 = \sqrt{m}(a_{31}x_1 + a_{32}x_2 + a_{33}x_3 + a_{34}x_4 + a_{35}x_5 + a_{36}x_6) \tag{4.177}$$

$$q_4 = \sqrt{m}(a_{41}x_1 + a_{42}x_2 + a_{43}x_3 + a_{44}x_4 + a_{45}x_5 + a_{46}x_6) \tag{4.178}$$

$$q_5 = \sqrt{m}(a_{51}x_1 + a_{52}x_2 + a_{53}x_3 + a_{54}x_4 + a_{55}x_5 + a_{56}x_6) \tag{4.179}$$

Table 4.4 Displacement of the atoms for $\lambda = 0$

Position of the atoms for $\lambda = 0$

$\lambda = 0$		a_1	a_2	a_3	a_4	a_5	a_6
0	a_1	$\frac{1}{\sqrt{3}}$	0	$\frac{1}{\sqrt{3}}$	0	$\frac{1}{\sqrt{3}}$	0
0	a_2	$\frac{1}{\sqrt{3}}$	0	$-\frac{1}{2\sqrt{3}}$	$\frac{1}{2}$	$-\frac{1}{2\sqrt{3}}$	$-\frac{1}{2}$
$\frac{3}{2}$	a_4	$\frac{1}{\sqrt{3}}$	0	$-\frac{1}{2\sqrt{3}}$	$-\frac{1}{2}$	$-\frac{1}{2\sqrt{3}}$	$\frac{1}{2}$
$\frac{3}{2}$	a_5	0	$\frac{1}{\sqrt{3}}$	$\frac{1}{2}$	$-\frac{1}{2\sqrt{3}}$	$-\frac{1}{2}$	$-\frac{1}{2\sqrt{3}}$
3	a_6	0	$\frac{1}{\sqrt{3}}$	$-\frac{1}{2}$	$-\frac{1}{\sqrt{3}}$	$\frac{1}{2}$	$-\frac{1}{2\sqrt{3}}$

$$q_6 = \sqrt{m}(a_{61}x_1 + a_{62}x_2 + a_{63}x_3 + a_{64}x_4 + a_{65}x_5 + a_{66}x_6) \tag{4.180}$$

This system of equations can be rewritten in matrix form as $Q = AX$, where

$$Q = \begin{pmatrix} q_1 \\ q_2 \\ q_3 \\ q_4 \\ q_5 \\ q_6 \end{pmatrix} = \sqrt{m} \begin{pmatrix} a_{11} & \cdots & a_{16} \\ \vdots & \ddots & \vdots \\ a_{61} & \cdots & a_{66} \end{pmatrix} \begin{pmatrix} x_1 \\ x_2 \\ x_3 \\ x_4 \\ x_5 \\ x_6 \end{pmatrix} \tag{4.181}$$

We have recorded the obtained results in Table 4.4, as reported in Tables 4.1, 4.2, and 4.3.

We can then correlate the coefficients in the following manner to establish a relationship between the coordinate systems.

$$\begin{pmatrix} a_{11} & a_{12} & a_{13} & a_{14} & a_{15} & a_{16} \\ a_{21} & a_{22} & a_{23} & a_{24} & a_{25} & a_{26} \\ a_{31} & a_{32} & a_{33} & a_{34} & a_{35} & a_{36} \\ a_{41} & a_{42} & a_{43} & a_{44} & a_{45} & a_{46} \\ a_{51} & a_{52} & a_{53} & a_{54} & a_{55} & a_{56} \\ a_{61} & a_{62} & a_{63} & a_{64} & a_{65} & a_{66} \end{pmatrix} = \begin{pmatrix} \frac{1}{\sqrt{3}} & 0 & \frac{1}{\sqrt{3}} & 0 & \frac{1}{\sqrt{3}} & 0 \\ \frac{1}{\sqrt{3}} & 0 & -\frac{1}{2\sqrt{3}} & \frac{1}{2} & -\frac{1}{2\sqrt{3}} & -\frac{1}{2} \\ 0 & \frac{1}{\sqrt{3}} & 0 & \frac{1}{\sqrt{3}} & 0 & \frac{1}{\sqrt{3}} \\ \frac{1}{\sqrt{3}} & 0 & -\frac{1}{2\sqrt{3}} & -\frac{1}{2} & -\frac{1}{2\sqrt{3}} & \frac{1}{2} \\ 0 & \frac{1}{\sqrt{3}} & \frac{1}{2} & -\frac{1}{2\sqrt{3}} & -\frac{1}{2} & -\frac{1}{2\sqrt{3}} \\ 0 & \frac{1}{\sqrt{3}} & -\frac{1}{2} & -\frac{1}{2\sqrt{3}} & \frac{1}{2} & -\frac{1}{2\sqrt{3}} \end{pmatrix} \tag{4.182}$$

Such that the generalized coordinates q_i in terms of the Cartesian coordinates x_i are related by the expression:

$$\begin{pmatrix} q_1 \\ q_2 \\ q_3 \\ q_4 \\ q_5 \\ q_6 \end{pmatrix} = \sqrt{m} \begin{pmatrix} \frac{1}{\sqrt{3}} & 0 & \frac{1}{\sqrt{3}} & 0 & \frac{1}{\sqrt{3}} & 0 \\ \frac{1}{\sqrt{3}} & 0 & -\frac{1}{2\sqrt{3}} & \frac{1}{2} & -\frac{1}{2\sqrt{3}} & -\frac{1}{2} \\ 0 & \frac{1}{\sqrt{3}} & 0 & \frac{1}{\sqrt{3}} & 0 & \frac{1}{\sqrt{3}} \\ \frac{1}{\sqrt{3}} & 0 & -\frac{1}{2\sqrt{3}} & -\frac{1}{2} & -\frac{1}{2\sqrt{3}} & \frac{1}{2} \\ 0 & \frac{1}{\sqrt{3}} & \frac{1}{2} & -\frac{1}{2\sqrt{3}} & -\frac{1}{2} & -\frac{1}{2\sqrt{3}} \\ 0 & \frac{1}{\sqrt{3}} & -\frac{1}{2} & -\frac{1}{2\sqrt{3}} & \frac{1}{2} & -\frac{1}{2\sqrt{3}} \end{pmatrix} \begin{pmatrix} x_1 \\ x_2 \\ x_3 \\ x_4 \\ x_5 \\ x_6 \end{pmatrix} \tag{4.183}$$

So for the generalized coordinates, we have functional expressions of the form $q_i = q_i(x_i)$, written as

$$q_1 = \sqrt{\frac{m}{3}}(x_1 + x_2 + x_3) \tag{4.184}$$

$$q_2 = \sqrt{\frac{m}{3}}\left(x_1 - \frac{1}{2}x_3 + \frac{\sqrt{3}}{2}x_4 - \frac{1}{2}x_5 - \frac{\sqrt{3}}{2}x_6\right) \tag{4.185}$$

$$q_3 = \sqrt{\frac{m}{3}}\,(x_2 + x_4 + x_6) \tag{4.186}$$

$$q_4 = \sqrt{\frac{m}{3}}\left(x_1 - \frac{1}{2}x_3 - \frac{\sqrt{3}}{2}x_4 - \frac{1}{2}x_5 + \frac{\sqrt{3}}{2}x_6\right) \tag{4.187}$$

$$q_5 = \sqrt{\frac{m}{3}}\left(x_2 + \frac{\sqrt{3}}{2}x_3 - \frac{1}{2}x_4 - \frac{\sqrt{3}}{2}x_5 - \frac{1}{2}x_6\right) \tag{4.188}$$

$$q_6 = \sqrt{\frac{m}{3}}\left(x_2 - \frac{\sqrt{3}}{2}x_3 - \frac{1}{2}x_4 + \frac{\sqrt{3}}{2}x_5 - \frac{1}{2}x_6\right) \tag{4.189}$$

In the development of the problem of the three masses coupled by springs that form the equilateral triangle, the following conventions for the coordinates of the atoms were used, such that the pairs (x_1, y_1), (x_2, y_2) and (x_3, y_3) constitute the positions of atoms 1, 2, and 3 respectively. Under this consideration, it is obtained:

$$\begin{pmatrix} x_1 \\ x_2 \\ x_3 \\ x_4 \\ x_5 \\ x_6 \end{pmatrix} = \begin{pmatrix} X_1 \\ Y_1 \\ X_2 \\ Y_2 \\ X_3 \\ Y_3 \end{pmatrix} \tag{4.190}$$

Therefore, the generalized coordinates q_i in terms of the position coordinates of atoms 1, 2, and 3 are

$$q_1 = \sqrt{\frac{m}{3}}(X_1 + X_2 + X_3) \tag{4.191}$$

$$q_2 = \sqrt{\frac{m}{3}}\left(X_1 - \frac{1}{2}X_2 + \frac{\sqrt{3}}{2}Y_2 - \frac{1}{2}X_3 - \frac{\sqrt{3}}{2}Y_3\right) \tag{4.192}$$

$$q_3 = \sqrt{\frac{m}{3}}\,(Y_1 + Y_2 + Y_3) \tag{4.193}$$

$$q_4 = \sqrt{\frac{m}{3}}\left(X_1 - \frac{1}{2}X_2 - \frac{\sqrt{3}}{2}Y_2 - \frac{1}{2}X_3 + \frac{\sqrt{3}}{2}Y_3\right) \tag{4.194}$$

$$q_5 = \sqrt{\frac{m}{3}}\left(Y_1 + \frac{\sqrt{3}}{2}X_2 - \frac{1}{2}Y_2 - \frac{\sqrt{3}}{2}X_3 - \frac{1}{2}Y_3\right) \tag{4.195}$$

$$q_6 = \sqrt{\frac{m}{3}}\left(Y_1 - \frac{\sqrt{3}}{2}X_2 - \frac{1}{2}Y_2 + \frac{\sqrt{3}}{2}X_3 - \frac{1}{2}Y_3\right) \tag{4.196}$$

The Cartesian coordinates x_i can be written in terms of the generalized coordinates q_i. This is achieved through the respective algebra, yielding:

$$X_1 = \frac{1}{\sqrt{3\,m}}(q_4 + q_1 + q_2) \tag{4.197}$$

$$Y_1 = \frac{1}{\sqrt{3\,m}}(q_6 + q_5 + q_3) \tag{4.198}$$

$$X_2 = \frac{1}{\sqrt{12\,m}}(-\sqrt{3}q_6 - q_4 + \sqrt{3}q_5 + 2q_1 - q_2) \tag{4.199}$$

$$Y_2 = \frac{1}{\sqrt{12\,m}}(-q_6 - \sqrt{3}q_4 + 2q_3 + \sqrt{3}q_2) \tag{4.200}$$

$$X_3 = \frac{1}{\sqrt{12\,m}}(\sqrt{3}q_6 - q_4 - \sqrt{3}q_5 + 2q_1 - q_2) \tag{4.201}$$

$$Y_3 = \frac{1}{\sqrt{12\,m}}(-q_6 + \sqrt{3}q_4 - q_5 + 2q_3 - \sqrt{3}q_2) \tag{4.202}$$

In general form, the coordinates X_i can be obtained through the inverse transformation, such that

$$Q = AX \rightarrow \; A^{-1}Q = A^{-1}AX \; \rightarrow X = A^{-1}Q \tag{4.203}$$

This indicates that in the coordinate transformation the matrix A must be non-singular for A^{-1} to exist.

4.4.2 Generalized Matrix Equation

Consider a system determined by the generalized coordinates that completely describe it. Let $q_k \rightarrow$ be the generalized coordinates, and if the system has n degrees of freedom, i.e., $k = 1, 2, 3...n$, we can write the following relationships:

$$x_i = x_i(q_k) \quad o \quad q_j = q_j(x_i) \tag{4.204}$$

$$T = \frac{1}{2}\sum_{jk} m_{jk}\dot{q}_j\dot{q}_k \tag{4.205}$$

$$\frac{\partial T}{\partial q_k} = 0 \quad \forall k \tag{4.206}$$

also

$$\left.\frac{\partial U}{\partial q_k}\right|_0 = 0 \quad k = 1, 2, 3...n \rightarrow Condicion\, de\, equilibrio \tag{4.207}$$

Since

$$U \equiv U(q_k) \tag{4.208}$$

We perform the expansion of the potential energy around the equilibrium point, that is, we perform the Taylor Series expansion:

$$\begin{aligned} U(x, y, x) = U(x_0, y_0, z_0) + \tfrac{\partial U}{\partial x}(x - x_0) + \tfrac{\partial U}{\partial y}(y - y_0) + \tfrac{\partial U}{\partial z}(z - z_0) + \\ \tfrac{\partial U}{\partial x \partial x}(x - x_0)^2 + \tfrac{\partial U}{\partial x \partial y}(x - x_0)(y - y_0) + ... \end{aligned} \tag{4.209}$$

$$U(q_1, q_2, q_3 ... q_n) = U_0 + \sum_k \left.\frac{\partial U}{\partial q_k}\right|_0 + \frac{1}{2}\sum_{j_1 k} \frac{\partial U}{\partial q_j}\frac{\partial U}{\partial q_k} q_j q_k + ... O(q_1, q_2, q_3 ... q_n) \tag{4.210}$$

$$Sea\ U_0 = 0\ \wedge\ \left.\frac{\partial U}{\partial q_k}\right|_0 = 0\ equilibrio \tag{4.211}$$

$$U(q_p) = \frac{1}{2}\sum_{j,k} A_{jk} q_j q_k\ con\ A_{jk} = \left.\frac{\partial^2 U}{\partial q_j \partial q_k}\right|_0 \tag{4.212}$$

If U has continuous second partial derivatives Si U tiene segundas derivadas parciales continuas

$$A_{jk} = A_{kj} \quad \Rightarrow\ Simetrica \tag{4.213}$$

Then

$$T = \frac{1}{2}\sum_{jk} m_{jk}\dot{q}_j\dot{q}_k \quad \wedge \quad U = \frac{1}{2}\sum_{j,k} A_{jk} q_j q_k \tag{4.214}$$

The quantities A_{jk} are numbers, however, m_{jk} can be functions of the coordinates.

$$m_{jk}(q_p) = m_{jk}(q_0) + \sum_l \left.\frac{\partial m_{jk}}{\partial q_l}\right|_0 q_l + ... \tag{4.215}$$

If m_{jk} is diagonal Si es diagonal

$$T = \frac{1}{2}\sum_r m_r \dot{q}_r^2 \quad m_{jk} = m_{kj} \quad \forall j = k \quad m_{jj} = m_r \tag{4.216}$$

If A_{jk} is diagonal, → The potential energy is the sum of the individual potentials.

The problem is to find a coordinate transformation that simultaneously diagonalizes both m_{jk} ∧ and A_{jk}

Such coordinates that are sought are called **Normal Coordinates** q_k, and they are the coordinates used in the Lagrange equations expressed as

$$\frac{\partial}{\partial t}\left(\frac{\partial T}{\partial \dot{q}_k}\right) - \frac{\partial T}{\partial q_k} = Q_k \tag{4.217}$$

with

$$\bar{F} = -\bar{\nabla}_{q_k} U \tag{4.218}$$

if

$$T \equiv T(\dot{q}_k) \quad \wedge \quad U \equiv U(q_k) \tag{4.219}$$

The Lagrange Equation will be

$$\frac{\partial}{\partial t}\left(\frac{\partial L}{\partial \dot{q}_k}\right) - \frac{\partial L}{\partial q_k} = 0 \tag{4.220}$$

Since $L = T - U$

$$L(\dot{q}_k, q_k) \Rightarrow \frac{\partial L}{\partial q_k} = \frac{\partial U}{\partial q_k} \quad \wedge \quad \frac{\partial L}{\partial \dot{q}_k} = -\frac{\partial T}{\partial \dot{q}_k} \tag{4.221}$$

$$\frac{\partial}{\partial t}\left(\frac{\partial T}{\partial \dot{q}_k}\right) - \frac{\partial U}{\partial q_k} = 0 \tag{4.222}$$

$$\frac{\partial U}{\partial q_k} = \frac{\partial}{\partial q_k}\left(\frac{1}{2}\sum_{jk} A_{jk} q_j q_k\right) = \frac{1}{2}\sum_{jk} A_{jk}\left(\frac{\partial q_j}{\partial q_k} q_k + q_j \frac{\partial q_k}{\partial q_k}\right) \tag{4.223}$$

$$\frac{\partial q_j}{\partial q_k} = \delta_{jk} \Rightarrow \frac{\partial U}{\partial q_k}\frac{1}{2}\left(\sum_{jk} A_{jk}(q_j + q_j) = \frac{1}{2}\sum_{j} A_{jk} q_j\right) \tag{4.224}$$

$$\frac{\partial U}{\partial q_k} = \frac{1}{2}\sum_{j} A_{jk} q_j \quad \wedge \quad \frac{\partial T}{\partial \dot{q}_k} = \frac{1}{2}\sum_{j} m_{jk} \dot{q}_j \tag{4.225}$$

$$\sum_{j}\left(m_{jk}\ddot{q}_j + A_{jk} q_j\right) = 0 \qquad j = 1, 2, 3..n \tag{4.226}$$

Equation 4.226 constitutes n differential equations whose solution is

$$q_j(t) = a_j e^{i(\omega t - \delta)} \tag{4.227}$$

Substituting into (4.226):

$$\sum_{j}\left(A_{jk} - \omega^2 m_{jk}\right) a_j = 0 \tag{4.228}$$

$$\left|A_{jk} - \omega^2 m_{jk}\right| = 0 \qquad j, k = 1, 2, 3...n \tag{4.229}$$

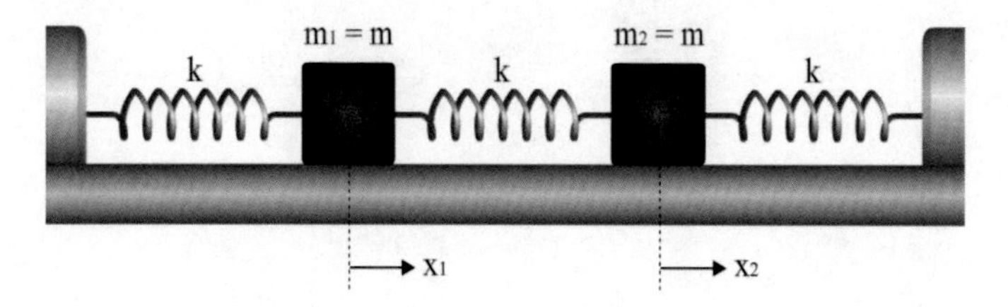

Fig. 4.14 System of 2 coupled masses attached to the walls

Example

Let us use the equations and methods of the generalized matrix technique to visualize the potentiality of the solution. Let there be two identical masses coupled by three springs, as shown in Fig. 4.14, where the additional constraints of the system due to the walls are depicted.

$$\boxed{T = \frac{1}{2}m\dot{x}_1^2 + \frac{1}{2}m\dot{x}_2^2}$$

$$U = \frac{1}{2}kx_1^2 + \frac{1}{2}k(x_2 - x_1) + \frac{1}{2}kx_2^2 \tag{4.230}$$

$$U = k\left(x_1^2 - x_2x_1 + x_2^2\right) \tag{4.231}$$

as

$$\boxed{A_{jk} = \left.\frac{\partial^2 U}{\partial q_j \partial q_k}\right|_0} \qquad A_{jk} = A_{kj} \tag{4.232}$$

$$q_1 = x_1 \qquad q_2 = x_2$$

$$A_{11} = \left.\frac{\partial^2 U}{\partial x_1 x_1}\right|_0 = \left.\frac{\partial^2 U}{\partial x_1^2}\right|_0 \qquad \boxed{A_{11} = 2k} \tag{4.233}$$

$$A_{12} = \left.\frac{\partial^2 U}{\partial x_1 \partial x_2}\right|_0 \tag{4.234}$$

$$\frac{\partial}{\partial x_1}(-kx_1 + 2kx_2)|_0 = -k = A_{21} \qquad \boxed{A_{21} = A_{12} = -k} \tag{4.235}$$

$$A_{22} = \left.\frac{\partial^2 U}{\partial x_2 x_2}\right|_0 = 2k \qquad \boxed{A_{22} = 2k} \tag{4.236}$$

$$T = \frac{1}{2}\sum_{jk} m_{jk}\dot{x}_j\dot{x}_k \tag{4.237}$$

$$T = \frac{1}{2}\left(m_{11}\dot{x}_1^2 + m_{22}\dot{x}_2^2 + 2m_{12}\dot{x}_1\dot{x}_2\right) \tag{4.238}$$

Comparing (4.4.2) and (4.238), it is obtained:

$$\boxed{m_{11} = m_{22} = m} \qquad \boxed{m_{12} = 0} \tag{4.239}$$

Setting up the determinant given by Eq. 4.232;

$$\begin{vmatrix} A_{11} - \omega^2 m_{11} & A_{12} - \omega^2 m_{12} \\ A_{21} - \omega^2 m_{21} & A_{22} - \omega^2 m_{22} \end{vmatrix} = 0 \tag{4.240}$$

$$\begin{vmatrix} 2k - \omega^2 m & -k \\ -k & 2k - \omega^2 m \end{vmatrix} = 0 \tag{4.241}$$

$$\left(2k - \omega^2 m\right)^2 - k^2 = 0 \;\Rightarrow\; 2k - \omega^2 m = \pm k \tag{4.242}$$

$$\omega^2 = \frac{2k \pm k}{m} \;\Rightarrow\; \omega_1 = \sqrt{\frac{k}{m}}, \quad \omega_2 = \sqrt{\frac{3k}{m}} \tag{4.243}$$

Example

Let's look at the following example, where two identical masses m are coupled by a single spring with elastic constant k. This is similar to the previous example but with a significant difference: the constraints of the system to the walls. See Fig. 4.15, where the coordinates x_1 and x_2 locate the positions of the masses.

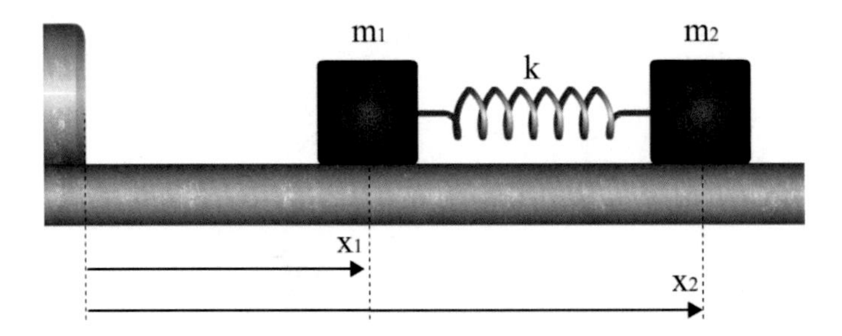

Fig. 4.15 System of 2 coupled masses

We define the potential energy of the spring as $U = \frac{1}{2}kx$, where x indicates the deformation of the spring, which for this example is defined by $x_2 - x_1$, so that

$$U = \frac{1}{2}k\left(x_2 - x_1\right)^2 = \frac{1}{2}k(x_2^2 - 2x_1x_2 + x_1^2) \tag{4.244}$$

Similarly, in Cartesian coordinates, the kinetic energy is

$$T = \frac{1}{2}m_1\dot{x}_1^2 + \frac{1}{2}m_2\dot{x}_2^2 \tag{4.245}$$

$$A_{jk} = \left.\frac{\partial^2 U}{\partial q_j \partial q_k}\right|_0 \quad \Rightarrow \quad A_{11} = \left.\frac{\partial^2 U}{\partial x_1^2}\right|_0 = k, \quad A_{12} = A_{21} = \left.\frac{\partial^2 U}{\partial x_1 x_2}\right|_0 = -k \tag{4.246}$$

$$A_{22} = k, \quad T = \frac{1}{2}\sum_{jk} m_{jk}\dot{x}_j\dot{x}_k = \frac{1}{2}\left(m_{11}\dot{x}_1^2 + m_{22}\dot{x}_2^2 + 2m_{12}\dot{x}_1\dot{x}_2\right) \tag{4.247}$$

$$m_{11} = m_1, m_{22} = m_2, m_{12} = m_{21} = 0$$

$$\begin{vmatrix} k - \omega^2 m_1 & -k \\ -k & k - \omega^2 m_2 \end{vmatrix} = 0 \tag{4.248}$$

$$\left(k - \omega^2 m_1\right)\left(k - \omega^2 m_2\right) - k^2 = 0 \tag{4.249}$$

$$k^2 - k\omega^2 m_2 - k\omega^2 m_1 + \omega^4 m_1 m_2 - k^2 = 0 \tag{4.250}$$

$$\omega^2\left(\omega^2 m_1 m_2 - k(m_1 + m_2)\right) = 0 \quad \Rightarrow \quad \omega = 0 \tag{4.251}$$

or

$$\omega^2 = \frac{k(m_1 + m_2)}{m_1 + m_2} \tag{4.252}$$

$$\boxed{\omega = \sqrt{\frac{k}{\mu}}} \quad \Rightarrow \quad \mu = \frac{m_1 m_2}{m_1 + m_2} \quad \Rightarrow \quad \frac{1}{\mu} = \frac{1}{m_1} + \frac{1}{m_2} \tag{4.253}$$

By solving the Schrödinger equation for two masses coupled to a spring, we obtain the energy:

$$E_n\left(n+\frac{1}{2}\right)\hbar\omega \Rightarrow E_n\left(n+\frac{1}{2}\right)\hbar\sqrt{\frac{k}{\mu}} \tag{4.254}$$

$$\Delta E = E_n - E_{n-1} = \hbar\sqrt{\frac{k}{\mu}}\left[\left(n+\frac{1}{2}\right)-\left(n-1+\frac{1}{2}\right)\right] = \hbar\sqrt{\frac{k}{\mu}} \tag{4.255}$$

$$\boxed{\Delta E = \hbar\sqrt{\frac{k}{\mu}}} \tag{4.256}$$

As we can see, the ground state energy is not zero and takes the value $E_1 = \left(\frac{1}{2}\right)\hbar\omega$, which is called the zero-point energy. This is the energy that the oscillators would have at absolute zero temperature. This zero-point energy is in accordance with the uncertainty principle.

Let is consider the following reasoning:

If the energy is zero, it means that the total energy is zero, hence, both kinetic and potential energy are zero. Under this consideration, it is inferred that momentum and position are well-defined, since $p = 0$ and $x = 0$. If the uncertainty principle is valid, this implies that the zero-point energy must necessarily be different from zero.

4.5 Other Coordinates

Normal coordinates, from a mathematical point of view, are the most simplified, but to define the Lagrangian it is necessary to know the explicit functional form of the potential energy. However, in the experimental process, the oscillation frequencies of the molecules and other data such as line width and behavior as a function of temperature are known, and the objective is to understand the bond strength, spatial geometry, respective group, and the environment in which the molecule is immersed. Normal coordinates allow for an elegant and mathematically rigorous treatment, but they do not intuitively visualize what happens in the molecule. There is a formalism of local modes, where displacement coordinates are defined according to the direction of the bonds and thus write the functional form of the potential, but the drawback is that only partial results are obtained. Another formalism introduces a complete set of coordinates, called internal coordinates, properly located within the molecule. Internal coordinates can be classified as:

(a) Stretching coordinates: Relate to the extension or elongation of the bond measured between the respective atoms of the molecule.

(b) Bending coordinates: Relate to the angle formed between the bonds.

Fig. 4.16 Tension

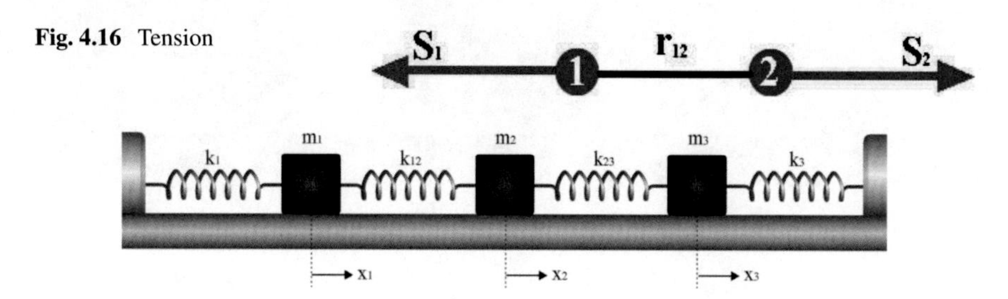

Fig. 4.17 System of 3 coupled masses with fixed ends

(c) Out-of-plane bending coordinates: Relate to the angle formed between a bond and the plane formed by two other different bonds.

(d) Torsion coordinates: Variation of the dihedral angle defined by three consecutive bonds.

It is possible to establish the relationship between normal coordinates and internal coordinates through the respective transformation (Fig. 4.16).

4.5.1 Concept of Phonons

The interaction between light and phonons in solids is considered, where phonons are vibrations of the atoms within the crystal structure that manifest in the infrared frequency region. It is to be remembered that emission and absorption in the visible and ultraviolet region are due to the bound electrons or those forming the bonds within the crystal. The main behavior of phonons can be explained in an understandable manner from the classical point of view, considering the use of dipolar oscillators. Initially, it allows us to understand how polar solids strongly absorb light in the infrared frequency band. It is also necessary to introduce the concepts of polaritons and polarons before considering the physics of inelastic scattering, where the Raman and Brillouin techniques are considered, complementing the information obtained by the Raman technique.

4.6 Proposed Problems

4.1 Obtain the normal coordinates and frequencies of the system consisting of three coupled masses with fixed ends, as shown in Fig. 4.17, using Lagrange's equations.

What are the frequencies if the masses are equal?

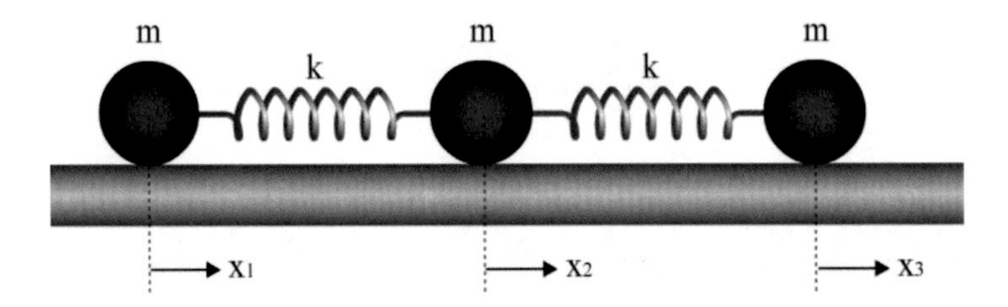

Fig. 4.18 3-mass coupled system

4.2 Obtain the normal coordinates and frequencies of the system consisting of two coupled masses, with one end fixed and the other end free. Use Lagrange's equations to find the respective solution.

What are the frequencies if the masses are equal?

4.3 Obtain the normal coordinates and frequencies of the system formed by three coupled masses with free ends, where the mass at the center differs from the masses at the ends (see Fig. 4.18). Use Lagrange's equations to find the respective solution.

How do the frequencies change if all the masses are equal?

References

1. Wells DA (1967) Theory and problems of lagrangian dynamics. McGraw-Hill
2. Marion JB, Stephen TT (2003) Classical dynamics of particles and systems, 5th ed. Brooks Cole
3. Wilson EB Jr, Decius JC, Cross PC (1980) Molecular vibrations: the theory of infrared and Raman vibrational spectra. Dover Publications
4. Herzberg G (1945) Infrared and Raman spectra of polyatomic molecules. D Van Nostrand

Chapter 5
Raman Spectroscopy

5.1 Theoretical Foundations of Raman Spectroscopy

In the physical process of the interaction of electromagnetic radiation with matter, the so-called scattering occurs, which is the deviation of light from its original direction of incidence on the material. This scattering process is described by a functional expression of the emission frequency as a function of the wave vector. The interaction of the electric field vector of the incident electromagnetic wave with the electrons of the material with which it interacts produces the scattering of the waves due to the oscillations of the electrons within the material, which in turn induces oscillating electric dipoles. These oscillating electric dipoles constitute new emission sources that re-emit radiation in all directions, with two basic types of scattering:

Elastic Scattering: It is the radiation emerging from the material where the emitted light frequency is equal to the incident light frequency, and it is known as Rayleigh scattering.

Inelastic Scattering: It is the radiation emerging from the material where the emitted light frequency is different from the incident light frequency.

Rayleigh scattering is the most probable, and it is due to this that we can observe objects. The efficiency in scattering is inversely proportional to the fourth power of the incident wavelength. Part of the light scattered inelastically is called Raman scattering. When the scattered light has lower energy than the light incident on the material, the effect is called Stokes Raman scattering, and in the opposite case, when the scattered light has higher energy than the incident light on the material, the effect is called anti-Stokes Raman scattering.

In this section, the classical and quantum models of the Raman effect will be obtained using the phenomenological representation of the harmonic oscillator subject to an external perturbation. The scattering of electromagnetic radiation with atomic motifs, represented by forced oscillators, will indicate that the model is

C. Vargas Hernandez, *Introduction to Raman Spectroscopy and Its Applications*,
https://doi.org/10.1007/978-3-031-77551-2_5

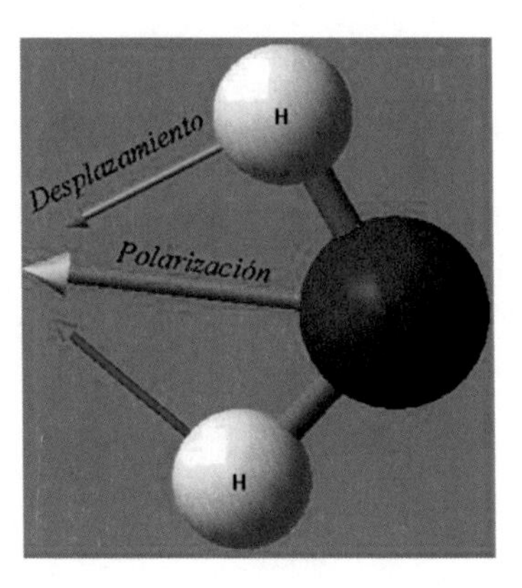

Fig. 5.1 Normal mode of vibration of the Water Molecule

acceptable and constitutes a good representation. Raman scattering is inelastic and dependent on the symmetry of the system being perturbed, but in the text, a deep explanation from the perspective of group theory will not be addressed; its mention will be superficial. Raman scattering is explained by the interaction of incident light with the electrons of the molecule within the material, and the energy is often not enough to excite these electrons to a higher electronic energy level, so its effect is to change the vibrational or rotational state of the molecule.

Within the broad phenomenology of Radiation-Matter interaction, we can mention the inelastic light scattering process within which the Raman process is found. An illustrative way to present the phenomenon is as follows: When light impacts the material, the molecules present in it interact with the photons, activating some of their normal modes of vibration, causing the molecule to oscillate in that specific mode. It is, of course, important to clarify that this is a primary model and that in this mentioned process, the rules of Physics that allow the phenomenon to happen are obeyed. There is not just one normal mode of vibration; molecules have several intrinsic or inherent normal modes of vibration of the material they constitute. For example, the water molecule, when considered isolated, has, according to quantum mechanical calculations, three inherent normal modes of vibration. Figure 5.1 illustrates one of them, which occurs at $1799\,cm^{-1}$.

For a molecule to exhibit the Raman effect, as It will see later, the incident light on it must induce a change in the dipole moment or a change in the molecular polarizability that can be visualized as a distribution of the electronic cloud.

5.2 Historical Review of the Raman Spectroscopy

In 1923, when Indian physicist Chandrasekhara Venkata Raman and his student were studying the scattering of light in alcohols, they observed a change in the color of the light passing through these solutions. They proceeded to carry out a filtering process of the scattered radiation to discriminate the Rayleigh component. This new radiation was successfully explained five years later, in 1928. For this contribution to science, he was awarded the Nobel Prize in Physics in 1930.

5.3 Classical Theory of the Raman Effect

When electromagnetic radiation, especially in the visible range, interacts with matter, it undergoes a scattering process, which consists of:

- The so-called Rayleigh scattering , characterized by high intensity and having the same frequency as the incident radiation. This means it is a process where the energy of the incident photon is equal to the energy of the emitted photon, and it is termed an elastic process.
- The so-called Raman scattering, which is very weak, is about 10^{-6} in comparison to the intensity of the incident radiation. One important characteristic is that the energies of the incident and emitted photons are not equal. This phenomenon is a type of inelastic scattering process.
  The frequency of Raman scattering obeys

$$v_0 \pm v_m \tag{5.1}$$

  where v_0 is the natural vibrational frequency of the molecule. $v_0 - v_m$ and $v_0 + v_m$ are called the Stokes and anti-Stokes lines, respectively. This means that Raman spectroscopy measures the molecular vibrational frequency, and its measurement is carried out using radiation in the UV-visible region.

Raman spectroscopy is an inelastic scattering process involving a photon. It entails the interaction between two particles: a photon and an electron belonging to a molecule. This interaction activates normal modes, such as vibration, rotation, or a vibro-rotational combination. During this process, energy conservation is maintained. If an incident photon with energy E_i interacts with the molecule, the molecule interacts with the photon through the electrons that constitute it, generating a new photon with energy E_f. Thus, by performing an energy balance, it can be inferred that the energy used to activate this normal mode of vibration is determined by

$$
\left|
\begin{array}{l}
E_i = \frac{hc}{\lambda_i} \\
E_f = \frac{hc}{\lambda_f} \\
E_i - E_f = \Omega \rightarrow \frac{hc}{\lambda_i} - \frac{hc}{\lambda_f} = \Omega \\
\Omega = hc\left(\frac{\lambda_f - \lambda_i}{\lambda_f \lambda_i}\right)
\end{array}
\right.
\tag{5.2}
$$

Just as there are many variations in absorption spectroscopy techniques, there are also additional techniques in Raman spectroscopy that utilize this inelastic scattering effect. In Raman scattering, two photons are involved: one incident photon that is absorbed and another photon that is emitted from the sample. This is a two-photon process occurring in a single step, where both Stokes Raman and anti-Stokes Raman scattering are possible. In the Stokes process, the emitted photon has lower energy, while in the anti-Stokes process, the emitted photon has higher energy.

Classical theories offer a close visualization of experimental behavior. In some cases, the approximation is adequate; in others, it is not as close but provides a very illustrative interpretation from a terminological point of view. To achieve this illustration, the Raman effect will be interpreted for the case of a diatomic molecule, highlighting one of the important characteristics, such as electrical polarization. In the radiation-matter interaction, radiation is considered in our case as an electromagnetic wave in the visible spectrum, represented by laser radiation. The electric field associated with the laser radiation is a function of time and can be represented by the expression:

$$
\mathbf{E} = \mathbf{E}_0 Cos(\omega t) \tag{5.3}
$$

With E_0 as the amplitude of oscillation of the incident electromagnetic field and ω as the frequency of the electromagnetic wave. When the molecule interacts with the electric field of the laser radiation, an electric dipole moment **P** is induced due to the redistribution of charge and is given by

$$
\mathbf{P} = \alpha \mathbf{E} = \alpha \mathbf{E}_0 Cos(\omega t) \tag{5.4}
$$

where α is the polarizability of the medium.

The electric field **E** of the light incident on a molecule induces a dipole moment in it, and the induced dipole moment μ is proportional to the electric field **E** that generates it, such that

$$
\mu = \alpha \mathbf{E} \tag{5.5}
$$

In complex systems such as solids, especially crystals, the concept of dipole moment μ is replaced by the macroscopic polarization of the medium **P** due to all the induced dipole moments in the material by the electric field. If the induced dipole moment **P** oscillates at a different frequency from the incident radiation, it results in an inelastic scattering called Raman scattering. If the photons scatter at lower frequencies than the excitation field, it is called Stokes scattering because part of their energy has been transferred to the interacting medium. When photons scatter at higher frequencies, it is called anti-Stokes scattering because the material medium gives up energy. Raman scattering has great potential as it provides both structural and compositional information about the sample under study. Using the principle of energy conservation and considering inelastic interactions where energy is either transferred or absorbed by the sample, it can be inferred that this quantum of energy has been used for the transition between two energy levels present in the sample. By measuring the energy difference between the incident and scattered photons, which is equal to the energy difference between the affected vibrational levels, each material shows a characteristic spectrum of frequencies due to its crystalline structure, bonds, types of atoms, impurities, etc. In other words, the Raman spectrum of each material represents a fingerprint that identifies the sample under study. The Raman frequencies emitted by the sample are independent of the excitation light. In scattering phenomena, approximately 0, 1% of the light incident on the sample is an elastic process and occurs with higher probability than the inelastic one, where only one in 10^6 or 10^7 incident photons is Raman. Within Raman scattering at room temperature, Stokes scattering is about 100 times greater than anti-Stokes because 99% of the molecules are in the lowest energy vibrational state, making it more likely that the sample absorbs energy to activate one of the normal vibrational modes and emits a photon of lower frequency.

The classical explanation of the Raman effect provides important information about the key characteristics of this technique. Considering the linear proportionality between the external electric field due to the incident radiation on the sample and the induced dipole moment as

$$\boldsymbol{\mu} = \overline{\overline{\alpha}} * \mathbf{E} \tag{5.6}$$

where α is the polarizability, which has a tensorial character due to anisotropy. When the dipole behavior is isotropic, α is a scalar. The linear approximation is valid if the internal fields of the excited molecule are greater than the electric field of the radiation that excites it. α measures the ability of the electron cloud of the excited molecule to be deformed by the external electric field.

Suppose the behavior of the electric field of the incident radiation is

$$\mathbf{E} = \mathbf{E}_0 e^{-i2\pi v_i t} \tag{5.7}$$

where v_i is the oscillation frequency of the field and E_0 is the maximum amplitude of the field. Additionally, the polarizability α is a function of the electronic distribution through the coordinate Q, which changes with the vibration of the molecule at a frequency v_m. Thus, the polarizability α can be expressed as a Taylor series expansion, as

$$\alpha = \alpha_0 + \left[\frac{\partial \alpha}{\partial Q}\right]_0 Q + \frac{1}{2}\left[\frac{\partial^2 \alpha}{\partial Q^2}\right]_0 Q^2 + \ldots \tag{5.8}$$

Taking the first-order approximation, we have

$$\alpha = \alpha_0 + \left[\frac{\partial \alpha}{\partial Q}\right]_0 Q \tag{5.9}$$

where α_0 is the polarizability at the equilibrium position. The variation of the coordinate Q with the frequency can be approximated as

$$Q = Q_0 Cos(2\pi v_m t) \tag{5.10}$$

where v_m is the natural frequency of the molecule, we obtain for the polarizability:

$$\alpha(t) = \alpha_0 + \left[\frac{\partial \alpha}{\partial Q}\right]_0 Q_0 Cos(2\pi v_m t) \tag{5.11}$$

We substitute this expression into the Eq. 5.6, to obtain

$$\boldsymbol{\mu} = \alpha \mathbf{E} \rightarrow \boldsymbol{\mu} = \left\{\alpha_0 + \left[\frac{\partial \alpha}{\partial Q}\right]_0 Q_0 Cos(2\pi v_m t)\right\} \mathbf{E}_0 Cos(2\pi v_i t) \tag{5.12}$$

$$\boldsymbol{\mu} = \alpha_0 \mathbf{E}_0 Cos(2\pi v_i t) + \left[\frac{\partial \alpha}{\partial Q}\right]_0 Q_0 Cos(2\pi v_m t) \mathbf{E}_0 Cos(2\pi v_i t) \tag{5.13}$$

Applying the trigonometric identity:

$$Cos(a)Cos(b) = \frac{1}{2}\left[Cos(a+b) + Cos(a-b)\right] \tag{5.14}$$

produces an expression with cross products of the form:

$$\boldsymbol{\mu} = \alpha_o \mathbf{E}_0 Cos\,(2\pi v_i t) + \frac{1}{2}\left(\frac{\partial \alpha}{\partial Q}\right)_0 Q_0 \mathbf{E}_0 \left[Cos\,(2\pi\,(v_i + v_m)\,t) + Cos\,(2\pi\,(v_i - v_m)\,t)\right] \tag{5.15}$$

Equation 5.15 shows that the dipole induced in the molecule due to the incident radiation with frequency v_i produces scattered radiation at the same frequency v_i, which constitutes the first term called elastic scattering. The second and third terms are the inelastic scattering, with frequencies $(v_i + v_m)$ and $(v_i - v_m)$, respectively, where v_m is the frequency of one of the molecule's normal modes of vibration. The photons scattered with lower frequency, that is, with frequency $(v_i - v_m)$ are called Raman Stokes scattering , and those scattered at a frequency of $(v_i + v_m)$ are called Raman anti-Stokes scattering. The values given by differences between v_i and the Stokes or anti-Stokes values are called Raman shifts, and their magnitudes correspond to the infrared region in the electromagnetic spectrum.

Although the explanation from the perspective of electromagnetic theory is incomplete, it provides useful insights into the Raman effect, including the following:

- The polarization and intensity of both Rayleigh and Raman scattering have a linear dependence on the intensity of the incident laser.
- Only vibrations that produce a change in polarizability ($\frac{\partial \alpha}{\partial Q} \neq 0$) lead to the Raman effect, which constitutes the first selection rule for Raman scattering.
- The Raman shift can be positive (Stokes) or negative (Anti-Stokes). The relative intensities of the two effects, Stokes and Anti-Stokes, depend on the population of the vibrational levels, which is determined by Boltzmann statistics as a function of temperature.
- $\frac{\partial \alpha}{\partial Q}$ varies according to the electronic configuration of the molecules and is different for each of the different normal modes of vibration.
- $\frac{\partial \alpha}{\partial Q}$ is much smaller than α_0, meaning the Raman effect is much smaller than Rayleigh scattering.

A clear example of the significance of polarizability in the intensity of the Raman signal can be seen in the $C = C$ and $C = O$ systems. For the $C = C$ system, when the vibration occurs, and the bonds stretch a strong polarization is produced due to the displacement of the electron clouds, resulting in a strong Raman signal associated with this normal mode of vibration. In contrast, for the $C = O$ system, the vibration produces a smaller displacement of the electron cloud and, therefore, a weaker Raman signal.

5.4 Selection Rules

As we will see later, group theory can demonstrate that if a molecule with a center of symmetry has Raman-active vibrations, these same vibrations will not appear in the infrared and vice versa. This constitutes a molecular selection rule, which can be used to determine whether a molecular vibration is Raman or infrared active. This rule can be understood by the fact that during the interaction between the photon and the molecule, the total angular momentum of the ground electronic state must be conserved, allowing only specific transitions. This rule indicates that symmetric molecular vibrations concerning the molecule's center of symmetry cannot be observed in infrared spectroscopy, and antisymmetric molecular vibrations cannot be observed in Raman spectroscopy. This is known as the mutual exclusion rule. It means that infrared absorption is only possible if there is a change in the molecule's dipole moment μ in a specific normal mode of vibration. Thus the infrared absorption band depends on the change in dipole moment μ during the vibration, such that

$$I_{IR} \propto \left[\frac{\partial \mu}{\partial Q}\right]^2 \tag{5.16}$$

In the Raman selection rule, a vibration is active if the polarizability α changes when the mode of vibration occurs, meaning:

$$\left[\frac{\partial \alpha}{\partial Q}\right]_0 \neq 0 \tag{5.17}$$

Thus, the intensity of the Raman-active band depends on the change in polarizability during the molecular vibration, i.e.,

$$I_{Raman} \propto \left[\frac{\partial \alpha}{\partial Q}\right]^2 \tag{5.18}$$

A useful consequence of this rule is in the study of polymers, where infrared spectroscopy is used to obtain detailed information about the functional groups of the polymer chain, and Raman spectroscopy is used to characterize the carbon chains that form the backbone of the polymers. Deciding whether a particular mode of vibration is Raman active, based on the variation of polarizability during molecular vibration, is only correct for diatomic or triatomic molecules. To determine if a vibration in molecules with more complex atomic arrangements is Raman active, group theory must be used, exploiting the possible symmetries present in the molecule. A vibration that does not significantly affect polarizability produces an extremely weak Raman band. Thus, highly polar bonds such as the $O-H$ bond generally produce very weak Raman scattering. Strong Raman scattering is due to bonds with electron clouds distributed over a relatively large region of space, such as double or triple bonds in carbon chains. The electron clouds of such bonds are easily deformable by the action of an external electric field. Bending or stretching such a bond substantially changes its electron density, resulting in a significant change in the induced dipole moment.

The energy difference between the incident photon on the sample and the scattered photon corresponds to the energy required to activate a normal mode of vibration, and this difference is independent of the wavelength of the incident light used as the source for the material analysis. This differs from the fact that the intensity of light scattering, including Raman scattering, depends on the wavelength of the incident light, with the scattering behavior being inversely proportional to λ^4 [1].

One of the important parameters associated with Raman intensity is the cross section σ_j; it has units of square centimeters per molecule. σ_j is related to $\frac{\partial \alpha}{\partial Q}$. Through theoretical treatments, it is possible to determine the intensity of the Raman signal I_R, which is related to polarizability by

$$I_R = \mu(v_0 \pm v_j)^4 \alpha_j^2 Q_j^2 \tag{5.19}$$

where μ is a constant. The above equation indicates that the intensity of Raman scattering varies with the fourth power of the observed frequency and shows a dependence on the laser frequency used. The factor v_j^4 is obtained through classical theoretical calculations of scattering by an oscillating induced dipole.

The above considerations allow us to establish differences between Raman scattering and infrared absorption spectroscopy, one of which relates to the possibility of producing both effects. In infrared spectroscopy, a change in the dipole moment is required for the normal mode of vibration to be allowed, whereas in Raman scattering, a change in polarizability is necessary. The infrared and Raman spectra differ both in the signal intensity and in the modes that can be visualized, consistent with the condition of activation of the normal mode of vibration. Another important characteristic is that when there are molecules with a center of inversion, the Raman and infrared modes are mutually exclusive. Regarding the techniques, it can be mentioned that Raman spectroscopy is a technique where a scattered signal is captured at different angles relative to the incident light source, whereas in infrared spectroscopy, the intensity of the absorbed signal is generally measured. Therefore, the choice of technique for a particular analysis depends on the analyte of interest, which is related to the selection rules.

5.5 Relation Between Stokes and Anti-Stokes Intensities

Consider an external electromagnetic field $\mathbf{E} = \mathbf{E}_0 e^{i2\omega_0 t}$ that induces a dipole moment $\boldsymbol{\mu}$ in the molecule, oscillating with the frequency ω_0 of the incident radiation. The classical electrodynamic theory for dipolar radiation proposed by Hertz indicates that an oscillating electric dipole emits radiation, with the time-averaged power radiated per unit solid angle expressed as

$$\left\langle \frac{dI}{d\Omega} \right\rangle = \left(\frac{\pi}{\varepsilon_0} \right)^2 \tilde{\upsilon}_0^4 \alpha^2 I_0 Sen\,(\theta) \quad (5.20)$$

where I_0 is the intensity of the incident radiation and θ is the angle between the field **E** and the observation direction. Applying this equation to Raman scattering, the Stokes and anti-Stokes processes are obtained as

$$\left\langle \frac{dI}{d\Omega} \right\rangle^{Stokes} = \left(\frac{\pi}{\varepsilon_0} \right)^2 (\tilde{\upsilon}_0 - \tilde{\upsilon}_m)^4 \alpha^2 I_0 Sen\,(\theta) \quad (5.21)$$

$$\left\langle \frac{dI}{d\Omega} \right\rangle^{Anti-Stokes} = \left(\frac{\pi}{\varepsilon_0} \right)^2 (\tilde{\upsilon}_0 + \tilde{\upsilon}_m)^4 \alpha^2 I_0 Sen\,(\theta) \quad (5.22)$$

Although the terminological explanation from the perspective of classical electrodynamic theory is acceptable for some macroscopic behaviors, it is not very adequate at the microscopic level and therefore does not explain some experimental observations that contradict the predictions. For example, if it establishes the ratio between the Stokes and anti-Stokes intensities, it gets

$$\frac{\left\langle \frac{dI}{d\Omega} \right\rangle^{Anti-Stokes}}{\left\langle \frac{dI}{d\Omega} \right\rangle^{Stokes}} = \frac{(\tilde{\upsilon}_0 + \tilde{\upsilon}_m)^4}{(\tilde{\upsilon}_0 - \tilde{\upsilon}_m)^4} > 1 \quad (5.23)$$

This does not agree with experimental behavior; hence, it is necessary to turn to quantum mechanics theory and the use of quantum statistics. The energetic position of the Stokes and anti-Stokes lines relative to the Rayleigh line is symmetrical, implying that the energy for the transition in both cases is the same. However, at room temperature, the Stokes line has a much higher intensity than the anti-Stokes line. As illustrated earlier, the Stokes line is generated when the incident photon induces a molecular transition from a lower state to a higher state, leaving the molecule in an excited state. In the anti-Stokes line, the transition occurs from a higher state to a lower state, meaning the molecule must be in an excited state and relax to a lower state, usually the ground state.

At room temperature, the number of molecules in the ground state is greater than the number of molecules in the excited state, implying a higher probability of the molecule undergoing a Stokes-type transition. Quantitatively, the relative population of molecules between the ground state and the excited state is determined by the Boltzmann distribution:

$$\frac{N_0}{N_i} = e^{-\frac{\Delta E}{kT}} \tag{5.24}$$

Where N_i is the number of molecules in the excited state with energy E_i, N_0 is the number of molecules in the ground state with energy E_0,k is the Boltzmann constant, T is the temperature in Kelvin, and ΔE is the energy difference between states E_i and E_0. At room temperature, approximately 99% of the molecules are in the ground state.

5.6 Raman Cross Section

Earlier, the dependence of both wavelength and the material's polarizability on the intensity of the Raman signal was discussed. One of the important parameters in the analytical treatment of signals is the cross section, as it has a strong dependence on the type of material being analyzed. The Raman cross section σ_j is proportional to the probability that an incident photon will be scattered as a Raman photon. In optical spectroscopy, one of the commonly used parameters is the molar absorptivity obtained from Beer's law, which represents the cross section for optical absorption.

Classical theory shows the relationship between the intensity of Raman scattering I_R, the cross section σ_j, and the intensity of the incident laser I_0 through

$$I_R = I_0 \sigma_j D dz \tag{5.25}$$

where D is the numerical density of scatterers (molecules per cubic centimeter) and dz is the optical path length of the laser within the sample. The frequency-independent cross section σ_j^0 can be defined as

$$\sigma_j^0 = \frac{\sigma_j}{(\bar{\upsilon}_0 - \bar{\upsilon}_j)^4} \tag{5.26}$$

where $\bar{\upsilon}_0 - \bar{\upsilon}_j$ are the wavenumbers expressed in cm^{-1}. Since modern signal acquisition instruments process the number of photons per second rather than power in watts, and calculations are more consistent with quantum mechanical treatment where the cross section is interpreted as a probability, for a photon-counting system, it can write

$$P_R = P_0 \sigma_j' D dz \tag{5.27}$$

where P_R and P_0 have units of photons per second, and σ_j' has a different frequency dependence than σ_j. Substituting σ_j as a function of ${\sigma_j}^0$ in the equation for I_R, and knowing that $I_R = P_R hc(\bar{v}_0 - \bar{v}_j)$ and $I_0 = P_0 hc\bar{v}_0$, we obtain the equation for the number of photons per second P_R:

$$P_R = P_0 {\sigma_j}^0 \bar{v}_0 (\bar{v}_0 - \bar{v}_j)^3 Ddz \tag{5.28}$$

with

$$\sigma_j' = \sigma_j^0 \bar{v}_0 (\bar{v}_0 - \bar{v}_j)^3 \tag{5.29}$$

Since the cross section σ_j includes scattering in all directions, the integral cross section must be determined in all directions and over all wavelength ranges of the Raman band being studied. This implies that the measurement of P_R requires collecting light in a solid angle of 4π steradians (sr) around the sample and integrating over the entire analyzed Raman band. In practice, due to the configuration of the measurement system, only a small range of solid angles is observed and in one or several directions of light scattering by the sample. Thus, for practical purposes, it is more useful to define a differential Raman cross section, $\beta = \frac{d\sigma_j}{d\Omega}$, where ω represents the solid angle used to collect the signal. The units of β are $cm^2 molecula^{-1} sr^{-1}$. The most appropriate and practical way to determine the Raman cross section is to quantitatively compare the Raman signal of a known or standardized sample with a cross section that serves as a normalization parameter against the unknown cross section. In the literature, β is determined by knowing the spectrometer's response function where the ratio of the peak areas of two bands under study is equal to the ratio of the cross sections.

In Raman intensity detection, there are other signals that are processed and compete in reducing the signal-to-noise ratio, including optical absorption and fluorescence. The cross section of these signals is 6 to 8 orders of magnitude greater than the Raman cross section, so many technical aspects and experimental designs focus on reducing the fluorescence signal.

5.7 Quantum Theory of the Raman Effect

The importance of understanding Raman spectroscopy lies in its wide range of applications as a material analysis technique, which is reflected in the increasing number of original articles presenting new applications with innovative methodologies. This progress is due to significant advances in instrumentation and optoelectronic devices, as well as the strengthening of modern optics and digital signal processing. The explanation of phenomena at the microscopic level for different states of matter is described by quantum mechanics theory. In the solid state, the behavior of atomic motifs, such as isolated molecules, can be studied and the entire solid can be considered by introducing average parameters, as in the case of crystal field theory. From this perspective, the isolated behavior of molecules is a good first approximation to infer their behavior within the solid, though collective behavior, which offers interesting aspects, is not accounted for. It will illustrate the quantum explanation of the Raman scattering effect by molecules.

Raman scattering involves a virtual state and two photons, one being the incident photon and the other the scattered photon. To establish the quantum explanation, it is necessary to employ second-order time-dependent perturbation theory. However, a pathway can be established where, if we know the coefficients Cp determined by first-order perturbation theory, according to the equation:

$$\sum_{n} C_p(t)e^{\frac{i(E_{pn})t}{\hbar}} V_{pn} = -\frac{\hbar}{i}\frac{dC_p(t)}{dt} \tag{5.30}$$

where, $\langle p| V(t) |n\rangle$ is the perturbation matrix; producing the transition from state p to n.

Assuming that the above equation is solved, we know the coefficients C_p, and with these coefficients, we find the terms C_m corresponding to the second-order approximation, i.e.,

$$\sum_{p} C_p^{(1)}(t)e^{\frac{i(E_{pm})t}{\hbar}} V_{pm} = -\frac{\hbar}{i}\frac{dC_m^2(t)}{dt} \tag{5.31}$$

where; $C_p^{(1)}$ and C_m^2 are the coefficients obtained from first- and second-order perturbation theory, respectively.

As the coefficients $C_p^{(1)}$, are written as

$$C_p(t) = \frac{V_{0if}}{2i\hbar}\left(\frac{1 - e^{i(\omega_{pi}+\omega)t}}{\omega_{pi}+\omega} - \frac{1 - e^{i(\omega_{pi}-\omega)t}}{\omega_{pi}-\omega}\right) \tag{5.32}$$

Substituting this value into Eq. 5.32 and after performing the algebra and respective simplifications, the probability of Raman scattering occurrence is obtained through the equation:

$$P_{i\to f} \propto \sum_{j} \frac{V_{ji}V_{fj}}{(\omega_{ji}+\omega) - (\omega_{ji}+\omega^{\otimes})} \tag{5.33}$$

where; $\omega^{\otimes} = \omega - \omega_{if}$, substituting this expression yields

$$P_{i\to f} \propto \sum_{j} \frac{V_{ji}V_{fj}}{\left(\omega_{ji}^2 - \omega^2\right)} \tag{5.34}$$

With the matrix elements $V_{ji} = \langle j| V(t) |i\rangle$ y $V_{fi} = \langle f| V(t) |i\rangle$. If V_{mn} are elements of the dipolar operator matrix such that $V_{mn} = \langle m| E\mu |n\rangle$, where E is the electric field and μ is the dipole moment, it means that the probability $P_{i\rightarrow f} \propto E^2$, and summing up yields polarizability as obtained in classical treatment.

5.8 Relation Between Symmetry, Selection Rules, and the Intensities of Stokes and Anti-Stokes Lines

Before delving into the basic fundamentals of group theory, it is important to visualize the relationships between the symmetries addressed through group theory and the selection rules that indicate when Raman transitions are possible and how the intensities of Stokes and anti-Stokes lines strongly depend on the matrix elements V_{mn}.

When discussing symmetry in molecules, it can typically be visualized geometrically for example: If a molecule is in a defined spatial position and a rotation is applied so that the new spatial distribution of the molecule is indistinguishable from the initial configuration, this action is called a symmetry operation. Symmetry operations are defined by a symmetry operator $\hat{R}$ and have the property of commuting with the system's Hamiltonian operator $\hat{H}$.

5.8.1 Example of a Raman Spectroscopy Calculation

Consider a diatomic molecule with vibrational modes that can be obtained by Raman spectroscopy. The molecule has a vibrational frequency υ_0 in its ground state. For the obtained Raman spectrum, if the observed scattered photon is $15,370cm^{-1}$, determine the vibrational frequency υ_0 of the molecule, given that the incident light frequency υ_{inc} is known to be 15,670 cm^-1 (638 nm). The Raman spectroscopy setup measures the scattered light to determine vibrational transitions.

Calculate the shift and intensity of Raman scattering. This example involves calculations and conceptual understanding relevant to Raman spectroscopy, including how to determine Raman shifts, intensities, and the effects of temperature on spectral measurements.

The Raman shift $\Delta\upsilon$ is defined as the difference between the incident and scattered light frequencies. If the incident light has a frequency υ_{inc}, The Raman shift $\Delta\upsilon$ is given by

$$\Delta\nu = \nu_{inc} - \nu_{scat} \rightarrow \Delta\nu = 15,670\,\text{cm}^{-1} - 15,370\,\text{cm}^{-1} = 300\,\text{cm}^{-1} \quad (5.35)$$

then

$$v_0 = \Delta \nu = 300\,\text{cm}^{-1} \tag{5.36}$$

This value indicates the vibrational frequency of the molecule.

The intensity of Raman scattering I_{Raman} is proportional to the square of the Raman polarizability derivative concerning the vibrational mode. Assume that the polarizability α is given by

$$\alpha = \alpha_0 + \alpha_1 \cos(2\pi v_0 t) \tag{5.37}$$

where α_0 and α_1 are constants. The intensity of Raman scattering I is proportional to

$$I \propto \left(\frac{\partial \alpha}{\partial v_0}\right)^2 \tag{5.38}$$

then

$$\frac{\partial \alpha}{\partial v_0} = -2\pi \alpha_1 t \sin(2\pi v_0 t) \tag{5.39}$$

Therefore,

$$I \propto [-2\pi \alpha_1 t \sin(2\pi v_0 t)]^2 \tag{5.40}$$

What is the effect of Temperature on Raman Shift?

With increasing temperature, the population of higher vibrational states increases according to the Boltzmann distribution. This effect can shift the observed Raman peak and change the intensity. Higher temperatures may also increase the width of the Raman peaks due to increased thermal motion.

Reference

1. Chantry GW, Gebbie HA, Helson C (1964) Nature 203:1052

Part II
SERS, Group Theory, and Symmetries

In this part, it is necessary to introduce variants of Raman spectroscopy whose objective is to increase the signal intensity, response, data collection time, sensitivity, and resolution of the captured signal. One of these variants uses the behavior of the plasmon and its generation. In addition to being used as a means of amplifying and selecting the Raman signal, these intermediaries are usually nanometric structures, such as metallic particles or metallic grids, to take advantage of the plasmonic effect and, through this interaction, obtain information about the normal vibration modes of the samples under study. These plasmonic matrices or nanoparticle systems are widely used to increase optical signals through the respective spectroscopic techniques.

One of the techniques discussed in Part II is the SERS theory, in which the theoretical foundations and theory of the plasmon are presented, with particular attention to the electromagnetic theory associated with it. Since the Raman signal enhancement factor in the samples is considerable, we decided to include in Part II the group theory and the symmetries associated with the different systems, with their respective applications to molecules.

Chapter 6
SERS Theory

6.1 Theoretical Foundations

A system containing a neutral gas composed of heavy ions and light electrons are called plasma. Such systems can be found, as a first approximation, in metals and doped semiconductors. In metals, the neutral gas consists of ions with a positive charge character that forms the crystal lattice, surrounded throughout their volume by nearly free electrons that circulate throughout the material, thus it can be considered a plasma. The oscillations of these systems from the perspective of quantum mechanics are called plasmons, as they are quantized collective oscillations. Furthermore, as we have already discussed in detail, the phenomenology of the quantization of vibrations and quantum oscillations of molecular systems are determined as inelastic scattering explained by Raman's theory. These two theories form a phenomenological set that intertwines to form surface-enhanced Raman spectroscopy (SERS) [1–3], which essentially involves amplifying the Raman signal by several orders of magnitude through the electromagnetic interaction of light with metallic particles or metallic surfaces.

6.2 SERS Theory

One of the major difficulties in optical spectroscopies is the signal-to-noise ratio. Due to this, the use of high-sensitivity amplifiers and detectors has increased significantly. However, the limitation is that in some cases, the background noise is also amplified, and in some techniques, these background signals are generated by electronic transitions that constitute fluorescence, which is very detrimental to the visualization of the Raman signal.

Around 1970, an interesting effect called Surface-enhanced Raman spectroscopy (SERS) was discovered. SERS is a spectroscopic technique where the Raman signal

C. Vargas Hernandez, *Introduction to Raman Spectroscopy and Its Applications*,
https://doi.org/10.1007/978-3-031-77551-2_6

is amplified due to surface phenomena occurring on metallic surfaces. This phenomenon led to the exploration and foundation of so-called metal-enhanced spectroscopic techniques. The SERS effect explained as an increase in the Raman signal from certain molecules in the presence of metallic nanostructures, is currently understood through two fundamental mechanisms:

- The electromagnetic model (EM)
- The chemical or charge transfer model (CT).

The exact mechanism of the Raman amplification effect (SERS) is still open to discussion. The two aforementioned theories differ in the mechanism producing the amplifying effect. The electromagnetic theory proposes the excitation of localized surface plasmons, while the chemical theory suggests the formation of complexes where charge transfer occurs. The latter is applicable if there are species that have formed a chemical bond with the nanoparticle surface, but it cannot always explain the observed signal increase through chemical bond formation alone. In contrast, the electromagnetic theory can be applied even when the sample is only physisorbed on the surface.

Both mechanisms contribute to the SERS effect, but the contribution of each depends on the system and the coupling between the analyzed molecules and the metallic nanostructures. Among the great advantages of the SERS technique, in addition to signal enhancement, are high sensitivity and selectivity. This allows for the use of small sample quantities and local environment analysis. The morphology of the metallic nanostructures used in SERS is important as it implies necessary coupling conditions with the incident radiation. The morphologies of metallic nanostructures can be controlled and designed through various chemical synthesis processes.

In the SERS spectroscopy technique, the signal enhancement is supported by the Raman signal increase, and the central parameter measuring this enhancement is the enhancement factor (EF). The goal is to theoretically or experimentally determine this factor, which spans a wide range of values from 10 to 10^{15}. This dispersion of values can be due to the not completely known structure of the nanoparticles as well as the strategy of calculating or measuring the parameter, generating difficulties when comparing theoretical predictions with experimental results. Therefore, it is important to clearly define the meaning and scope of this enhancement factor, considering that we refer to Stokes Raman scattering, where the average Raman intensity is mainly proportional to both the intensity of the incident laser power and the effective cross section of the molecule. Accordingly, it is expected that the SERS intensity for a specific normal vibration mode of a studied molecule is proportional to both the incident laser intensity and the cross section of the interacting molecules, related to the enhancement factor (EF) of the Raman signal. The EF constitutes the essence of the SERS technique, making it a powerful analysis and diagnostic tool. One way to obtain a tentative value of the EF in the first approximation could be to measure the sample under identical conditions to measure the Raman signal and the SERS signal at a specific control peak, associated with a normal vibration mode of the studied molecule. The EF is generally considered to be the contribution of two factors: the electromagnetic enhancement factor F_{EM}, which in most cases is the main contribution, and the chemical enhancement factor CE.

6.3 Plasmons and Electromagnetic Theory

The theory is framed for low frequencies of electromagnetic excitation radiation as an effect within linear optics, which depends on the first power of the incident intensity I_0. Of course, it is worth noting that nonlinear optical processes also produce an increase in the Raman signal, but this chapter will only address the linear process.

The increase in the intensity of the Raman signal through the SERS effect is associated with the relationship between the excitation electric field E_{exc} (i.e., the field of the incident electromagnetic wave) and the average local electric field radiated by the surface of the particle within the study sample E_m. Let g be the parameter that determines this increase, such that we can express this relationship by

$$E_m = gE_{exc} \tag{6.1}$$

Similarly, let E_R be the magnitude of the Raman field radiated by the molecule under study. This field is associated with the light scattered by that molecule. Let α_R be the amplification factor of the Raman field in the molecule when radiation from the metallic particle with a field intensity E_m is incident on it. Therefore, we have

$$E_R = \alpha_R E_m \tag{6.2}$$

Using the same reasoning, we have

$$E_{SERS} = \acute{g}E_R \tag{6.3}$$

where E_{SERS} is the magnitude of the SERS field radiated by the metallic particle when a field of magnitude E_R from the molecule is incident on it. The factor $\acute{g}$ is the proportionality between these two fields.

By using the above equations, we obtain the functional expression between the magnitude of the SERS field and the incident excitation field, defined as

$$E_{SERS} = \acute{g}E_R = \acute{g}\alpha_R E_m = \acute{g}\alpha_R g E_{exc} \tag{6.4}$$

The intensity of the electromagnetic radiation I is proportional to the square of the magnitude of the electric field, i.e.,

$$I = |E|^2 \tag{6.5}$$

Therefore, the SERS intensity I_{SERS} is

$$I_{SERS} = |E_{SERS}|^2 = \left|\acute{g}\alpha_R g E_{exc}\right|^2 \tag{6.6}$$

Thus,

$$I_{SERS} = \left|g\acute{g}\right|^2 |\alpha_R|^2 |E_{exc}|^2 \tag{6.7}$$

For low excitation frequencies where we can approximate $g \cong \acute{g}$, the SERS intensity can be increased by a factor proportional to the fourth power of the increase in the local incident field, i.e.,

$$|E_L|^4 = |g|^4 \tag{6.8}$$

For high-frequency Raman modes, the SERS intensity can be a complicated function due to the resonance properties of the plasmons present in the metallic particles with the frequencies of both the incident and scattered Raman waves.

According to the above, we can define the SERS signal enhancement factor G as the ratio between the intensity of the Raman signal scattered in the presence of the metallic particle and in the absence of the metallic particle, using the equation:

$$G = \left|\frac{\alpha_R}{\alpha_{R_0}}\right| \left|g\acute{g}\right|^2 \tag{6.9}$$

where α_{R_0} is the Raman polarizability of the isolated molecule.

In the presence of metallic nanoparticles, the intensity of the Raman signal is enhanced by several orders of magnitude, making it possible to detect a single molecule. This is why SERS spectroscopy has become an extremely sensitive detection technique and is of vital importance when the amount of material available for analysis is very small, such that normal techniques like Infrared or Raman itself produce weak signals or cannot record them. For this reason, SERS is considered one of the few physical phenomena currently framed within nanoscience.

6.3.1 Electromagnetic Theory in Metals

The theory of the interaction between electromagnetic radiation and metals leads to the phenomenology known as plasmonics. All electromagnetic theory is based on Maxwell's equations. The equations that explain the behavior of electric and magnetic fields and their relationship with moving charges are called Maxwell's equations and can be expressed in integral and differential forms, as follows:

$$\left|\begin{array}{l} \oint \mathbf{E}\cdot d\mathbf{s} = \frac{q}{\varepsilon_0} \\ \oint \mathbf{B}\cdot d\mathbf{s} = 0 \\ \oint \mathbf{E}\cdot d\mathbf{l} = -\frac{\partial}{\partial t}\int_S \mathbf{B}\cdot d\mathbf{s} \\ \oint \mathbf{B}\cdot d\mathbf{l} = \mu_0 I + \mu_0\varepsilon_0\frac{\partial}{\partial t}\int_S \mathbf{E}\cdot d\mathbf{s} \end{array}\right. \tag{6.10}$$

$$\left|\begin{array}{l} \nabla\cdot\mathbf{E} = \frac{\rho}{\varepsilon_0} \\ \nabla\cdot\mathbf{B} = 0 \\ \nabla\times\mathbf{E} = -\frac{\partial\mathbf{B}}{\partial t} \\ \nabla\times\mathbf{B} = \mu_0\mathbf{J} + \mu_0\varepsilon_0\frac{\partial\mathbf{E}}{\partial t} \end{array}\right. \tag{6.11}$$

Let's derive the wave equation for empty space, that is, for a space free of electric charges and currents. In such a case, the differential equations for the electric field are

$$\nabla\cdot\mathbf{E} = 0 \tag{6.12}$$

$$\nabla\times\mathbf{B} = \mu_0\varepsilon_0\frac{\partial\mathbf{E}}{\partial t} \tag{6.13}$$

Knowing the following vector identity:

$$\nabla\times\nabla\times\mathbf{A} = \nabla\left(\nabla\cdot\mathbf{A}\right) - \nabla^2\mathbf{A} \tag{6.14}$$

Applying it to the curl of the electric field, we have

$$\nabla\times\nabla\times\mathbf{E} = \nabla\left(\nabla\cdot\mathbf{E}\right) - \nabla^2\mathbf{E} \tag{6.15}$$

$$\nabla\times\nabla\times\mathbf{E} = -\frac{\partial\nabla\times\mathbf{B}}{\partial t} = -\mu_0\varepsilon_0\frac{\partial}{\partial t}\left(\frac{\partial\mathbf{E}}{\partial t}\right) \tag{6.16}$$

$$-\nabla^2 \mathbf{E} = -\mu_0 \varepsilon_0 \frac{\partial}{\partial t}\left(\frac{\partial \mathbf{E}}{\partial t}\right) \tag{6.17}$$

where the Laplacian is defined as

$$\nabla^2 = \frac{\partial^2}{\partial x^2} + \frac{\partial^2}{\partial y^2} + \frac{\partial^2}{\partial z^2} \tag{6.18}$$

Then,

$$\frac{\partial^2 \mathbf{E}}{\partial x^2} + \mu_0 \varepsilon_0 \frac{\partial^2 \mathbf{E}}{\partial t^2} = 0 \tag{6.19}$$

Comparing with the one-dimensional wave equation, we have

$$\frac{\partial^2 A}{\partial x^2} + \frac{1}{v^2}\frac{\partial^2 A}{\partial t^2} = 0 \tag{6.20}$$

Therefore, we have for the speed of electromagnetic waves

$$v = \sqrt{\frac{1}{\mu_0 \varepsilon_0}} \tag{6.21}$$

The parameter that measures the behavior of the material to electromagnetic fields is the dielectric constant, and it is obtained from the constitutive equations of the electromagnetic fields, given by

$$D(K, \omega) = \varepsilon_0 \varepsilon(K, \omega) E(K, \omega) \tag{6.22}$$

$$J(K, \omega) = \sigma(K, \omega) E(K, \omega) \tag{6.23}$$

where K is the wave vector of the electromagnetic wave and J is the electric current density.

As

$$D = \varepsilon_0 E + P, \quad J = \frac{\partial P}{\partial t} \tag{6.24}$$

In the Fourier frequency domain, we have

$$\frac{\partial}{\partial t} \to -i\omega \tag{6.25}$$

Therefore,

$$J = \frac{\partial P}{\partial t} = -i\omega P = \sigma E \rightarrow P = i\frac{\sigma E}{\omega} \tag{6.26}$$

Then,

$$D = \varepsilon_0 E + P, \rightarrow \varepsilon_0 \varepsilon(K,\omega) E(K,\omega) = \varepsilon_0 E(K,\omega) + i\frac{\sigma(K,\omega)}{\omega} E(K,\omega) \tag{6.27}$$

Thus,

$$\varepsilon_0 \varepsilon(K,\omega) = \varepsilon_0 + i\frac{\sigma(K,\omega)}{\omega} \tag{6.28}$$

where finally we have

$$\varepsilon(K,\omega) = 1 + i\frac{\sigma(K,\omega)}{\varepsilon_0 \omega} \tag{6.29}$$

$$\varepsilon = \varepsilon_1 + i\varepsilon_2, \cdots \sigma = \sigma_1 + i\sigma_1 \tag{6.30}$$

$$\tilde{n} = n(\omega) + ik(\omega), \quad \tilde{n} = \sqrt{\varepsilon} \tag{6.31}$$

$$\varepsilon_1 = n^2 - k^2, \quad \varepsilon_2 = 2nk \tag{6.32}$$

$$n^2 = \frac{\varepsilon_1}{2} + \frac{1}{2}\sqrt{\varepsilon_1^2 + \varepsilon_2^2}, \quad k = \frac{\varepsilon_2}{2n} \tag{6.33}$$

where, σ, ϵ, k and n are the electrical conductivity, the dielectric constant, the extinction coefficient, and the refractive index, respectively.

Another behavior of materials concerning electromagnetic waves is the surface absorption of the wave, which is governed by Beer's law, defined as

$$I = I_0 e^{-\alpha x}, \quad \alpha = \frac{2k(\omega)\omega}{c}, \quad \delta = \frac{1}{\alpha} \tag{6.34}$$

where α is the absorption coefficient and δ the skin depth.

6.3.2 Free Electron Gas Model

Let's consider the free electron model in metals moving in the presence of an external electric field and the positive ions constituting the material, thus forming a plasma. Let the electron have mass m. In the $\hat{x}$-direction, the forces acting are the external excitation force; $\mathbf{F} = \hat{x}qE_0\cos(\omega t)$ and the viscous force; $\mathbf{F}_{vis\cos a} = b\dot{x}$.

$$b\dot{x} + qE_0\cos(\omega t) = m\ddot{x} \tag{6.35}$$

$$m\ddot{x} - b\dot{x} = -qE_0\cos(\omega t) \tag{6.36}$$

$$\ddot{x} - \frac{b}{m}\dot{x} = -\frac{qE_0}{m}\cos(\omega t) \tag{6.37}$$

$$\ddot{x} - \Gamma\dot{x} = -\frac{qE_0}{m}e^{i\omega t} \tag{6.38}$$

Let the solution of the differential equation be

$$x(t) = Ae^{-i(\omega t - \phi)} \tag{6.39}$$

Thus, substituting into the previous equation, we obtain

$$\dot{x} = -i\omega Ae^{i(\omega t-\phi)}, \quad \ddot{x} = -\omega^2 Ae^{i(\omega t-\phi)} \tag{6.40}$$

$$-\omega^2 Ae^{i(\omega t-\phi)} + i\omega\Gamma Ae^{i(\omega t-\phi)} = -\frac{qE_0}{m}e^{i\omega t} \tag{6.41}$$

$$\omega^2 A - i\omega\Gamma A = \frac{qE_0}{m}e^{i\phi} \tag{6.42}$$

By equating the real and imaginary parts, we have

$$\frac{qE_0}{m}\cos\phi = \omega^2 A \tag{6.43}$$

$$\frac{qE_0}{m}sen\phi = -\omega\Gamma A \tag{6.44}$$

Therefore, the phase angle ϕ can be obtained as

$$\tan\phi = -\frac{\Gamma}{\omega} \tag{6.45}$$

$$\left(\frac{qE_0}{m}\cos\phi\right)^2 + \left(\frac{qE_0}{m}sen\phi\right)^2 = \left[\omega^2 A\right]^2 + [\omega\Gamma A]^2 \tag{6.46}$$

The amplitude $A(\omega)$ is obtained by dividing Eqs. 5.10 by 5.9

$$A(\omega) = \frac{{qE_0}/{m}}{\left[\left(\omega^2\right)^2 + (\omega\Gamma)^2\right]^{1/2}} \tag{6.47}$$

Therefore, the equation describing the motion of the plasma system with damping and forced by an external excitation with frequency ω is

$$x(t) = \frac{{qE_0}/{m}}{\left[\left(\omega^2\right)^2 + (\omega\Gamma)^2\right]^{1/2}}\cos(\omega t - \phi) \tag{6.48}$$

We define the polarization of the system as

$$P = -nqx \tag{6.49}$$

Then,

$$P(t) = \frac{{nq^2E_0}/{m}}{\left[\left(\omega^2\right)^2 + (\omega\Gamma)^2\right]^{1/2}}\cos(\omega t - \phi) \tag{6.50}$$

Another way to write it is using Eq. 6.41

$$-\omega^2 Ae^{i(\omega t-\phi)} + i\omega\Gamma Ae^{i(\omega t-\phi)} = -\frac{qE_0}{m}e^{i\omega t} \tag{6.51}$$

$$\omega^2 x(t) - i\omega\Gamma x(t) = \frac{q}{m}E(t) \tag{6.52}$$

Luego.

$$x(t) = \frac{q}{m(\omega^2 + i\omega\Gamma)}E(t) \tag{6.53}$$

Luego,

$$P = -\frac{nq^2}{m(\omega^2 + i\omega\Gamma)}E(t) \tag{6.54}$$

Since the dielectric displacement in a material is defined as

$$\bar{D} = \varepsilon_0\bar{E} + +\bar{P} \tag{6.55}$$

Then,

$$\bar{D} = \varepsilon_0\left[1 - \frac{nq^2/m\varepsilon_0}{(\omega^2 + i\omega\Gamma)}\right]\bar{E}(t) \tag{6.56}$$

$$\bar{D} = \varepsilon_0\left[1 - \frac{\omega_p^2}{(\omega^2 + i\omega\Gamma)}\right]\bar{E}(t) \tag{6.57}$$

where we have defined the plasma frequency of the system as

$$\omega_p^2 = \frac{nq^2}{m\varepsilon_0} \tag{6.58}$$

Furthermore,

$$\bar{D}(t) = \varepsilon_0\varepsilon\bar{E}(t) \tag{6.59}$$

It is obtained that

$$\varepsilon = \left(1 - \frac{\omega_p^2}{(\omega^2 + i\omega\Gamma)}\right) \tag{6.60}$$

It can also relate γ with the relaxation time of the system associated with the time between collisions,

$$\Gamma = \frac{1}{\tau} \tag{6.61}$$

The relaxation time in metals is on the order of $10^{-14}s$ seconds with damping factors $\gamma = 100\,\text{THz}$.

By multiplying Eq. 6.60 by its complex conjugate, It is obtained:

$$\varepsilon = \left(1 - \frac{\omega_p^2}{(\omega^2 + \Gamma^2)}\right) + i\left(\frac{\Gamma\omega_p^2}{\omega(\omega^2 + \Gamma^2)}\right) \tag{6.62}$$

Since ϵ is a complex function written by Eq. 6.29, and comparing with (6.62), we have

$$\varepsilon_1 = 1 - \frac{\omega_p^2 \tau^2}{(\omega^2 \tau^2 + 1)} \tag{6.63}$$

$$\varepsilon_2 = \frac{\omega_p^2 \tau}{\omega(\omega^2 \tau^2 + 1)} \tag{6.64}$$

Using Eq. 6.61, we obtain the dielectric constant as a function of relaxation time, as

$$\varepsilon_1 = 1 - \frac{\omega_p^2 \tau^2}{(\omega^2 \tau^2 + 1)}, \quad \varepsilon_2 = \frac{\omega_p^2 \tau}{\omega(\omega^2 \tau^2 + 1)} \tag{6.65}$$

In nonlinear electrodynamics, the dielectric constant ϵ and conductivity σ are functions of the electric field E and are expressed as

$$\varepsilon = \frac{\varepsilon_0}{\left[1 + \frac{1}{b^2}(c^2B^2 - E^2)^{1/2}\right]}, \quad \sigma = \mu_0\left(1 + \frac{1}{b^2}(c^2B^2 - E^2)^{1/2}\right) \tag{6.66}$$

6.3.2.1 High-Frequency Regime, $\omega \gg 10^{14}$ Hz

In this frequency range, we can say that the material retains its metallic character as we normally know it. If $\omega < \omega_p$ and $\tau = 10^{-14}s$, consider the visible range, for example, $\lambda = 400nm$, this implies that

$$\omega = \frac{2\pi c}{\lambda} \approx 5x10^{15} \tag{6.67}$$

Then, $\omega\tau = 50$, we can consider relaxation times less than or equal to 10^{-14}, such that

$$\omega\tau \gg 1 \tag{6.68}$$

Therefore, Eq. 6.63 leads to

$$\varepsilon_1 = 1 - \frac{\omega_p^2\tau^2}{(\omega^2\tau^2+1)} \approx 1 - \frac{\omega_p^2\tau^2}{\omega^2\tau^2} = 1 - \frac{\omega_p^2}{\omega^2} \tag{6.69}$$

Thus, we have

$$\varepsilon_1 = 1 - \frac{\omega_p^2}{\omega^2} \tag{6.70}$$

$$\varepsilon_2 = \frac{\omega_p^2\tau}{\omega(\omega^2\tau^2+1)} \approx \varepsilon_2 = \frac{\omega_p^2\tau}{\omega(\omega^2\tau^2)} \to 0 \tag{6.71}$$

This indicates that for frequencies below ultraviolet, the material is radiative, allowing the wave to propagate within it, with damping $\varepsilon_2 \to 0$. Care must be taken with the approximation in Eq. 6.71, depending on the range of ω_p, since:

$$\omega_p = \sqrt{\frac{nq^2}{m\varepsilon_0}}, \quad \omega_p \approx 6x10^{15}\,Hz \tag{6.72}$$

Therefore, as in the previous calculation in Eq. 6.67, $\varepsilon_2 = -0,04 \to 0$

$$\varepsilon = \varepsilon_1 + i\varepsilon_2 \approx \varepsilon_1 \tag{6.73}$$

The wave propagates within the metal and is minimally absorbed. The condition of $\omega < \omega_p$ and at high frequencies ensures that the wave propagates within the metal without absorption.

6.3.2.2 Low-Frequency Regime, $\omega \ll 10^{14}$ Hz

For the equivalent low-frequency or long-wavelength regime, such as infrared wavelengths or greater, $\omega \sim 10^{12}F$ Hz, we have

$$\omega\tau \approx 10^{12}x10^{-14} \approx 10^{-2}$$

$$\omega\tau \ll 1 \tag{6.74}$$

In this case, Eq. 6.63 leads to

$$\varepsilon_1 = 1 - \frac{\omega_p^2\tau^2}{(\omega^2\tau^2+1)} \approx 1 - \omega_p^2\tau^2 \tag{6.75}$$

$$\varepsilon_2 = \frac{\omega_p^2 \tau}{\omega(\omega^2 \tau^2 + 1)} \approx \varepsilon_2 = \frac{\omega_p^2 \tau}{\omega} \quad (6.76)$$

For our case $\omega_p \approx 6x10^{15} Hz$, in this case $\varepsilon_1 \ll 0$ and $\varepsilon_2 = 3.2x10^5$, which leads to the conclusion that

$$\varepsilon_2 \gg \varepsilon_1 \quad \rightarrow \quad \varepsilon \approx \varepsilon_2 \quad (6.77)$$

Additionally, considering Eq. 6.33, we have

$$n^2 = \frac{\varepsilon_1}{2} + \frac{1}{2}\sqrt{\varepsilon_1^2 + \varepsilon_2^2} \approx \frac{\varepsilon_1}{2} + \frac{\varepsilon_2}{2} \approx \frac{\varepsilon_2}{2} \quad (6.78)$$

That is

$$n \approx \sqrt{\frac{\varepsilon_2}{2}} \quad (6.79)$$

And written in terms of the plasma frequency and the field frequency using Eq. 6.77, we have

$$n = \sqrt{\frac{\omega_p^2 \tau}{2\omega}} \quad (6.80)$$

In this region, metals are absorptive and heat up due to the absorption of infrared or longer wavelengths.

6.3.2.3 Conductivity in Metals

The conductivity σ in metals is a function dependent on frequency, and its behavior in DC is defined as

$$\sigma_0 = \frac{nq^2 \tau}{m} \quad (6.81)$$

Thus, using Eq. 6.58, we have

$$\sigma_0 = \omega_p^2 \tau \varepsilon_0 \quad (6.82)$$

Using Eqs. 6.34, 6.30, and considering high frequencies as per (6.77), we have

$$\tilde{n} = k \approx \sqrt{\frac{\varepsilon_2}{2}} \tag{6.83}$$

Using equation (6.80), we get

$$k = \sqrt{\frac{\omega_p^2 \tau}{2\omega}} \tag{6.84}$$

Thus, the absorption coefficient is

$$\alpha(\omega) = \sqrt{\frac{2\omega_p^2 \tau \omega}{c^2}} \tag{6.85}$$

Using Eqs. (6.82) and (6.21), we get

$$\alpha(\omega) = \sqrt{2\sigma_0 \omega \mu_0} \tag{6.86}$$

And using Eq. (6.34), we obtain the skin depth:

$$\delta = \frac{1}{\sqrt{2\sigma_0 \omega \mu_0}} \tag{6.87}$$

The penetration of electromagnetic fields experimentally decays by 67% at distances of approximately $\delta = 100$ nm.

We can infer that the dielectric function in a metal consists of a real part ϵ_1 and an imaginary part ϵ_2. The attenuation of the electromagnetic wave in the metal, associated with the dissipation of energy as heat, is related to this imaginary term. As demonstrated previously, at high frequencies in the ultraviolet range or higher, absorption decreases, tending to zero, and the metal behaves as transparent. In the infrared frequency range or lower, this absorption increases, heating the metal.

6.3.3 Localized Surface Plasmons

To understand the behavior of localized plasmons, consider a particle of size a immersed in a medium m interacting with an electromagnetic field. We assume a homogeneous and isotropic sphere, and let $\bar{E} = E_0 \hat{z}$ be the electric field in an isotropic and non-absorptive medium with a dielectric constant ϵ_m, and let $\epsilon_N P(\omega)$ be the dielectric constant for the nanoparticle (NP).

Due to the azimuthal symmetry of the problem, the general solution is of the form:

$$\Phi(r,\theta) = \sum_{l=0}^{\infty} \left[A_l r^l + B_l r^{-(l+1)}\right] P_l(\cos\theta) \tag{6.88}$$

where $P_l(\cos\theta)$ are the Legendre polynomials of order l, and the parameters A_l and B_l are obtained from the boundary conditions.

6.3.3.1 Conducting Spherical Nanoparticle

Considering the conducting sphere, we can recall from Gauss's law that the electric field inside the conductor is zero, so

$$E(r) = 0, \quad \forall r < a \tag{6.89}$$

Then,

$$E = -\nabla\Phi \tag{6.90}$$

Therefore,

$$\Phi \equiv cte \tag{6.91}$$

Let us take the particular case where $\phi = 0$, meaning our sphere is grounded. Additionally, since the potential is finite, then

$$\Phi = 0, \quad \forall r < a \tag{6.92}$$

Therefore, Eq. 6.88 shows us that $B_l = 0$ for all l, as we must ensure that the potential is finite at $r = 0$, and we obtain

$$\Phi(r,\theta) = \sum_{l=0}^{\infty} \left[A_l r^l\right] P_l(\cos\theta), \quad \forall r < a \tag{6.93}$$

$$\Phi_{II}(r,\theta) = \sum_{l=0}^{\infty} \left[C_l r^l + D_l r^{-(l+1)}\right] P_l(\cos\theta), \quad \forall r > a \tag{6.94}$$

We have the following additional condition at infinity,

$$E(r,\theta)|_{r\to\infty} \to -E_0\hat{z} \tag{6.95}$$

Using Eq. 6.90, we obtain for the potential:

$$\Phi(r,\theta)|_{r\to\infty} \to -E_0 z = -E_0 r\cos\theta \tag{6.96}$$

But we know that

$$P_0(\cos\theta) = 1, \quad P_1(\cos\theta) = \cos\theta, \quad P_2(\cos\theta) = \frac{1}{2}(3\cos^2\theta - 1) \tag{6.97}$$

which indicates that we can write

$$\Phi(r,\theta)|_{r\to\infty} \to -E_0 r P_1(\cos\theta) \tag{6.98}$$

By using Eq. 6.94 and expanding the terms,

$$\begin{aligned} &\Phi(r,\theta)|_{r\to\infty} \to C_0 P_0 + D_0 r^{-1} P_0 + C_1 r P_1 + D_1 r^{-2} P_1 + \\ &\sum_{l\neq 0,1}^{\infty} \left[C_l r^l + D_l r^{-(l+1)}\right] P_l(\cos\theta) \to -E_0 r P_1(\cos\theta) \end{aligned} \tag{6.99}$$

which indicates that $C_1 = -E_0$, and $C_l = 0 \;\; \forall l \neq 1$. Therefore, for the potential function outside the sphere, we have

$$\Phi_{II}(r,\theta) = -E_0 r P_1(\cos\theta) + \sum_{l\neq 0,1}^{\infty} \left[D_l r^{-(l+1)}\right] P_l(\cos\theta), \quad \forall r > a \tag{6.100}$$

Applying the continuity conditions at the interface or surface of the sphere, which separates the sphere itself from the surrounding environment, we must consider that the electric field is also continuous. The normal and tangential components of the electric field are continuous in regions I and II, the internal and external regions of the sphere, respectively. We express this continuity as

$$E_{IT}(a) = E_{IIT}(a), \quad D_{IN}(a) = D_{IIN}(a) \tag{6.101}$$

where T and N stand for tangential and normal components, respectively. Since the Laplacian in spherical coordinates is expressed as

$$\bar{\nabla} = \hat{r}\frac{\partial}{\partial r} + \hat{\theta}\frac{1}{r}\frac{\partial}{\partial\theta} + \hat{\varphi}\frac{1}{rsen\theta}\frac{\partial}{\partial\varphi} \tag{6.102}$$

Considering Eq. 6.90 and taking into account the azimuthal symmetry (no variation with ϕ), the nabla operator becomes

$$\bar{\nabla} = \hat{r}\frac{\partial}{\partial r} + \hat{\theta}\frac{1}{r}\frac{\partial}{\partial \theta} \tag{6.103}$$

Thus, the normal and tangential components of the field are

$$\bar{\nabla}_N = \hat{r}\frac{\partial}{\partial r}, \quad \bar{\nabla}_T = \hat{\theta}\frac{1}{r}\frac{\partial}{\partial \theta} \tag{6.104}$$

therefore,

$$E_T = -\frac{1}{r}\frac{\partial \Phi}{\partial \theta}, \quad D_N = \varepsilon\frac{\partial \Phi}{\partial r} \tag{6.105}$$

The continuity conditions of Eq. 6.101 indicate that

$$E_{IT}(a) = E_{IIT}(a) \rightarrow -\frac{1}{a}\frac{\partial \Phi_I}{\partial \theta} = -\frac{1}{a}\frac{\partial \Phi_{II}}{\partial \theta} \tag{6.106}$$

then we have

$$\frac{\partial \Phi_I}{\partial \theta} = \frac{\partial \Phi_{II}}{\partial \theta} \tag{6.107}$$

The other continuity condition (6.101) indicates that

$$\varepsilon \left.\frac{\partial \Phi_I}{\partial r}\right|_{r=a} = \varepsilon_m \left.\frac{\partial \Phi_{II}}{\partial \theta}\right|_{r=a} \tag{6.108}$$

The problem to solve for the conducting sphere in the presence of an electric field in the z-direction is summarized by determining the parameters A_l and D_l, subject to the following continuity conditions,

$$\left\| \begin{array}{l} \Phi_I(r,\theta) = \sum\limits_{l=0}^{\infty} \left[A_l r^l\right] P_l(\cos\theta), \quad \forall r < a \\ \\ \Phi_{II}(r,\theta) = -E_0 r P_1(\cos\theta) + \sum\limits_{l=0}^{\infty} \left[D_l r^{-(l+1)}\right] P_l(\cos\theta), \quad \forall r > a \\ \\ -\frac{1}{a}\frac{\partial \Phi_I}{\partial \theta}\Big|_{r=a} = -\frac{1}{a}\frac{\partial \Phi_{II}}{\partial \theta}\Big|_{r=a}, \qquad \varepsilon \frac{\partial \Phi_I}{\partial r}\Big|_{r=a} = \varepsilon_m \frac{\partial \Phi_{II}}{\partial \theta}\Big|_{r=a} \end{array} \right. \tag{6.109}$$

we will substitute the boundary conditions into the equations, that is, we solve the system of Eq. 6.109

$$-\sum_{l=0}^{\infty} \left[A_l a^l\right] \frac{\partial}{\partial \theta} P_l(\cos\theta) = +E_0 a \frac{\partial}{\partial \theta} P_1(\cos\theta) - \sum_{l=0}^{\infty} \left[D_l a^{-(l+1)}\right] \frac{\partial}{\partial \theta} P_l(\cos\theta) \tag{6.110}$$

where we have used Eq. 6.97, we rewrite the equation by taking the first term out of the summation:

$$\begin{array}{l} -A_1 a^1 \frac{\partial}{\partial \theta} P_1(\cos\theta) - \sum\limits_{l\neq 1}^{\infty} \left[A_l a^l\right] \frac{\partial}{\partial \theta} P_l(\cos\theta) = E_0 a \frac{\partial}{\partial \theta} P_1(\cos\theta) - \\ \\ -D_1 a^{-2} \frac{\partial}{\partial \theta} P_1(\cos\theta) + \sum\limits_{l\neq 1}^{\infty} \left[D_l a^{-(l+1)}\right] \frac{\partial}{\partial \theta} P_l(\cos\theta) \end{array} \tag{6.111}$$

By equating the coefficients of the polynomials and considering that the Legendre polynomials are orthogonal and satisfy

$$si, \quad \sum_{l=0}^{\infty} C_l P_l(\cos\theta), \quad \rightarrow C_l = 0 \tag{6.112}$$

Accordingly, we have

$$-A_1 a^1 = E_0 a - D_1 a^{-2}, \quad \rightarrow A_1 = -E_0 + \frac{D_1}{a^3} \tag{6.113}$$

And applying Eq. 6.112 for the remaining terms, we have

$$-A_l a^l + D_l a^{-(l+1)} = 0, \quad \rightarrow A_l = \frac{D_1}{a^{2l+1}}, \quad \forall l \neq 1 \tag{6.114}$$

For the second continuity condition, we have

$$-\varepsilon\frac{\partial}{\partial r}\left[\sum_{l=0}^{\infty} A_l r^l P_l(\cos\theta)\right]\Bigg|_{r=a} = -\varepsilon_m\frac{\partial}{\partial r}\left[-E_0 r P_1(\cos\theta) + \sum_{l=0}^{\infty}\left[D_l r^{-(l+1)}\right] P_l(\cos\theta)\right]\Bigg|_{r=a} \tag{6.115}$$

$$-\varepsilon\left[\sum_{l=0}^{\infty} P_l(\cos\theta)\frac{\partial}{\partial r}A_l r^l\right]\Bigg|_{r=a} = \varepsilon_m E_0 P_1(\cos\theta)\frac{\partial}{\partial r}r - \sum_{l=0}^{\infty} P_l(\cos\theta)\frac{\partial}{\partial r}\left[D_l r^{-(l+1)}\right]\Bigg|_{r=a} \tag{6.116}$$

Deriving

$$-\varepsilon\sum_{l=0}^{\infty} P_l(\cos\theta)A_l l r^{l-1}\Bigg|_{r=a} = \varepsilon_m E_0 P_1(\cos\theta) + \sum_{l=0}^{\infty} P_l(\cos\theta)(l+1)D_l r^{-l-2}\Bigg|_{r=a} \tag{6.117}$$

Evaluating at a and extracting the first term from the summations, we have

$$-\varepsilon P_1(\cos\theta)A_1 - \varepsilon\sum_{l\neq 1}^{\infty} A_l l a^{l-1} P_l(\cos\theta) = \varepsilon_m E_0 P_1(\cos\theta) + 2\varepsilon_m P_1(\cos\theta)D_1 a^{-3} +$$

$$+\varepsilon_m\sum_{l=0}^{\infty} P_l(\cos\theta)(l+1)D_l a^{-l-2} \tag{6.118}$$

By equating the coefficients of the first term and using condition (6.112), we have

$$\begin{aligned} -\varepsilon P_1(\cos\theta)A_1 &= \varepsilon_m E_0 P_1(\cos\theta) + 2\varepsilon_m P_1(\cos\theta)D_1 a^{-3}, \\ -\varepsilon A_1 &= \varepsilon_m E_0 + 2\varepsilon_m D_1 a^{-3} \end{aligned} \tag{6.119}$$

Then we have

$$A_1 = -\frac{\varepsilon_m}{\varepsilon}\left(E_0 + \frac{2D_1}{a^3}\right) \tag{6.120}$$

For the remaining coefficients different from one, we have

$$-\varepsilon A_l l a^{l-1} = \varepsilon_m (l+1) D_l a^{-l-2} \tag{6.121}$$

Then we have

$$A_l = -\frac{\varepsilon_m (l+1)}{\varepsilon l a^{2l+1}} D_l, \quad \rightarrow \forall l \neq 1 \tag{6.122}$$

To find the values of A_1, D_1 and the A_l, D_l, we equate Eqs. 6.113 with 6.120 and 6.114 with 6.122, respectively.

$$-E_0 + \frac{D_1}{a^3} = -\frac{\varepsilon_m}{\varepsilon}\left(E_0 + \frac{2D_1}{a^3}\right), \quad \rightarrow D_1 = \left(\frac{\varepsilon - \varepsilon_m}{2\varepsilon_m + \varepsilon}\right) a^3 E_0 \tag{6.123}$$

and for A_1

$$A_1 = -E_0 + \frac{D_1}{a^3} = -E_0 + \frac{1}{a^3}\left(\frac{\varepsilon - \varepsilon_m}{2\varepsilon_m + \varepsilon}\right) a^3 E_0 \tag{6.124}$$

$$A_1 = -\left(\frac{3\varepsilon_m}{2\varepsilon_m + \varepsilon}\right) E_0 \tag{6.125}$$

For the D_l, we equate (6.114) with (6.122).

$$\frac{D_1}{a^{2l+1}} = -\frac{\varepsilon_m (l+1)}{\varepsilon l a^{2l+1}} D_l \tag{6.126}$$

Then,

$$D_l\left(1 + \frac{\varepsilon_m (l+1)}{\varepsilon l}\right) = 0, \quad \rightarrow D_l = 0 \quad \forall l \neq 1 \tag{6.127}$$

Therefore, the solution to find the potential at any point of a conducting sphere in the presence of a uniform electric field is

$$\begin{Vmatrix} \Phi_I(r,\theta) = A_1 r P_1(\cos\theta), \quad \forall r < a \\ \Phi_{II}(r,\theta) = -E_0 r P_1(\cos\theta) + D_1 r^{-2} P_1(\cos\theta), \quad \forall r > a \end{Vmatrix} \tag{6.128}$$

Considering that $P_1 \cos(\theta) = \cos(\theta)$, the solution is

$$\left\| \begin{aligned} &\Phi_I(r,\theta) = -\left(\tfrac{3\varepsilon_m}{2\varepsilon_m+\varepsilon}\right) E_0 r \cos(\theta), \quad \forall r < a \\ &\Phi_{II}(r,\theta) = -E_0 r \cos(\theta) + \left(\tfrac{\varepsilon-\varepsilon_m}{2\varepsilon_m+\varepsilon}\right) \tfrac{a^3 E_0}{r^2} \cos(\theta), \quad \forall r > a \end{aligned} \right. \tag{6.129}$$

The electric potential of a system can be written using a multipole expansion as follows:

$$\Phi(r,\theta) = \Phi_M(r,\theta) + \Phi_D(r,\theta) + \Phi_Q(r,\theta) \tag{6.130}$$

where M, D, and Q are the labels for the monopole, dipole, and quadrupole terms, respectively.

The first two are defined as

$$\Phi_M(r,\theta) = \frac{Q}{4\pi\varepsilon_0 r}, \quad \Phi_D(r,\theta) = \frac{\bar{P}\cdot\bar{r}}{4\pi\varepsilon_0\varepsilon_m r^3} \tag{6.131}$$

Knowing that

$$\bar{P} = \varepsilon_0\varepsilon_m\alpha\bar{E}_0 \tag{6.132}$$

where α is the polarizability of the medium. Comparing the dipole term for Φ_{II} in Eqs. 6.129 with 6.131, we have

$$\frac{\bar{P}\cdot\bar{r}}{4\pi\varepsilon_0\varepsilon_m r^3} = \left(\frac{\varepsilon-\varepsilon_m}{2\varepsilon_m+\varepsilon}\right)\frac{a^3 E_0}{r^2} \tag{6.133}$$

we get

$$P = 4\pi\varepsilon_0\varepsilon_m\left(\frac{\varepsilon-\varepsilon_m}{2\varepsilon_m+\varepsilon}\right)a^3 E_0 \tag{6.134}$$

By comparing Eqs. 6.132 with 6.134, we get the polarizability,

$$\alpha = 4\pi\left(\frac{\varepsilon-\varepsilon_m}{2\varepsilon_m+\varepsilon}\right)a^3 \tag{6.135}$$

Equation 6.135 is the key result in the interaction process. If the polarizability tends to zero, then the dipole moment P tends to zero, indicating that the material is non-polarizable, which is important in Raman spectroscopy.

As seen before, for high frequencies $\omega \geq 10^{14}$ Hz, the dielectric constant has a predominant real part,

$$\varepsilon_1 = 1 - \frac{\omega_p^2}{\omega^2} \approx \varepsilon, \quad \rightarrow \varepsilon = 1 - \frac{\omega_p^2}{\omega^2} = -2\varepsilon_m \tag{6.136}$$

thus at resonance Eq. 6.135, implies that $2\varepsilon_m + \varepsilon \rightarrow 0$, then

$$\frac{\omega_p^2}{\omega^2} = 1 + 2\varepsilon_m \tag{6.137}$$

Condition (6.138) is called the Frohlich condition, associated with dipolar surface plasmons of metallic nanoparticles.

$$\varepsilon = -2\varepsilon_m \tag{6.138}$$

If the medium is air $\epsilon = 1$, then in the resonance Eq. 6.137:

$$\omega = \frac{\omega_p}{\sqrt{3}} \tag{6.139}$$

The resonance shifts to the red when ϵ_m increases. It is important to note that the resonance of the absorption Eq. 6.135 is limited and restricted because the imaginary part can be small but not necessarily zero, preventing the absorption coefficient from tending to infinity.

The electric field is obtained from:

$$\bar{E} = -\bar{\nabla}\Phi = -\left(\hat{r}\frac{\partial}{\partial r} + \hat{\theta}\frac{1}{r}\frac{\partial}{\partial \theta} + \hat{\phi}\frac{1}{rsen\theta}\frac{\partial}{\partial r}\right)\Phi \tag{6.140}$$

By applying Eqs. 6.140 to 6.129, we have

$$\left\| \begin{array}{l} \bar{E}_I(r,\theta) = \left(\frac{3\varepsilon_m}{2\varepsilon_m+\varepsilon}\right)\bar{E}_0, \quad \forall r < a \\ \\ \bar{E}_{II}(r,\theta) = \bar{E}_0 + \frac{1}{4\pi\varepsilon_0}\frac{3(\bar{p}\cdot\hat{r})\hat{r}-\bar{p}}{r^3} - \frac{1}{3\varepsilon_0}\bar{p}\delta^3(\bar{r}), \quad \forall r > a \end{array} \right. \tag{6.141}$$

This field is of vital importance because, in Mie's theory, it is predicted that the field of a sphere in the presence of an external electric field is the same as when the particle has a size smaller than $\lambda/20$. This field at any point is equal to the superposition of the dipole field and the external field that excites the dipole. The resonance of the absorption coefficient α implies an increase in both the internal field and the dipolar field. This result is very important in applications of metallic nanoparticles in optical devices and sensors.

References

1. Moskovits M (1985) Surface-enhanced spectroscopy? Rev Mod Phys 57:783
2. Wokaun A (1985) Surface-enhancement of optical fields: mechanisms and applications. Mol Phys 56:1
3. Stockman M (2006) Electromagnetic theory of SERS, topic in applied physics. Surface-Enhanced Raman Scattering, vol 103. Springer

Chapter 7
Group Theory

The theory of groups plays a crucial role in spectroscopy, particularly in understanding and analyzing the vibrational spectra of molecules [1, 2]. Why is group theory important in spectroscopy? Here are several aspects:

- **Classification and Determination of Molecular Symmetry** Group theory provides a systematic and structured method for classifying the symmetry of molecules based on the spatial distribution of their components. By determining the point group of a molecule, we can predict which vibrational modes are active in infrared (IR) and Raman spectroscopy. This classification simplifies the analysis of complex spectra by identifying symmetry-related properties of the molecule and informing us about the appropriate techniques for obtaining these properties.
- **Selection Rules Applied to Molecules** Group theory helps derive selection rules for the energetic transitions between molecular states obtained from spectroscopic techniques. These rules dictate whether a particular vibrational mode is IR or Raman active. For example:
  - **IR Activity**: A vibrational mode is IR active if it involves a change in the dipole moment of the molecule.
  - **Raman Activity**: A vibrational mode is Raman active if it involves a change in the polarizability of the molecule. By using character tables and symmetry elements from group theory, we can determine which vibrational modes satisfy these conditions.
- **Reduction of Vibrational Modes**: Group theory allows us to decompose or reduce the vibrational modes of a molecule into irreducible representations, significantly simplifying experimental and theoretical studies. This reduction helps identify the number and types of vibrational modes (e.g., stretching, bending) and their symmetry properties, as well as assign these modes to the respective peaks obtained in spectroscopic techniques.

C. Vargas Hernandez, *Introduction to Raman Spectroscopy and Its Applications*,
https://doi.org/10.1007/978-3-031-77551-2_7

- **Understanding the Energy Degeneracy of Molecular Levels or Transitions**: Group theory explains the degeneracy of vibrational modes, showing how many times a particular energy level appears. In quantum mechanics, a degenerate level is one that has multiple states assigned to the same energy. This is crucial for understanding the splitting of spectral lines and predicting intensity patterns in spectra.
- **Prediction of spectroscopic patterns obtained by respective techniques**: By knowing the symmetry properties of a molecule, group theory helps to predict the overall pattern of the spectrum. For example, it can predict the number of peaks in a Raman or IR spectrum and their relative intensities based on the symmetry of the vibrational modes.
- **Correlation Diagrams**: Group theory facilitates the use of correlation diagrams to track how molecular orbitals transform under different symmetry operations. This is essential in understanding electronic transitions and how they couple with vibrational modes, influencing the spectra.
- **Simplifying Complex Spectra**: For larger or more complex molecules or crystals, group theory provides tools to break down the problem into smaller, more manageable parts by analyzing the symmetry of subunits or repeating patterns. This simplification is invaluable for interpreting complex vibrational spectra.

In summary, group theory is a powerful mathematical tool that enhances our understanding of molecular vibrations and rotations. It provides a clear and structured approach to predicting and interpreting molecular spectra. This is indispensable in the field of spectroscopy, as it complements and expands our knowledge of molecular symmetries and their interactions with electromagnetic radiation.

7.1 Group Theory

A group is defined as a set of elements $A, B, C, ...$, such that its elements satisfy the following properties under a binary operation:

$$G = \{A, B, C, ...\}$$

$$\begin{aligned}
&(1) \quad \forall x, y \in G \quad x * y = z \ / \ z \in G \\
&(2) \quad \forall x, y, z \in G \quad x * (y * z) = (x * y) * z \\
&(3) \quad \forall x \in G \quad \exists E \ / \ xE = Ex = x \\
&(4) \quad \forall x \in G \quad \exists x^{-1} \quad xx^{-1} = x^{-1}x = E
\end{aligned}$$

If the group's operation satisfies the commutative law, it is called a commutative group or Abelian group.

In group theory, it is necessary to define what a group means. A set of elements, such as A, B, C,..., that have common relations forms what it will call a multiplication group. The multiplication between its elements must satisfy the following rules:

- The product of two elements in the group generates a third element. If this third element is within the set that forms the group, the set is said to be closed under the multiplication rule established for the group.
- When there are three elements and the multiplication rule is invariant to the order in which the multiplication is performed, that is, (AB)C = A(BC), the associative law is said to be hold.
- The set of elements that make up the group must contain an element, which it will call E or identity, such that when multiplied by each of the elements of the group, the value of those elements does not change, that is, it leaves them invariant: EA = AE = A.
- The set of elements that make up the group must contain the inverse element A^{-1} for each of its elements, such that multiplying an element by its inverse generates the identity element E, that is, $A^{-1}A = E$.

A finite group is one that has a finite number of elements h. This number h represents the order of the group. In addition to the above properties, if all the elements of the set satisfy AB = BA, the group is said to be commutative or Abelian.

7.2 Representation of Geometric Transformations by Matrices

The set of symmetry elements of a molecule can be described by the following five types of operations:

- **Identity (E):** This is the simplest symmetry operation, where the molecule is left unchanged. Every molecule possesses this operation, as doing nothing to the molecule (leaving it as it is) is always a valid operation.
- **Rotation** (C_n): This operation involves rotating the molecule around an axis by a certain angle. The angle is typically a fraction of 360° (e.g., 180° for a C_2 axis). If the molecule is unchanged after the rotation, this rotation axis is a symmetry element.
- **Reflection** (σ)**:** This operation reflects the molecule through a mirror plane. If the molecule looks the same after the reflection, it possesses that mirror plane as a symmetry element. The mirror planes can be classified as vertical (σ_V), horizontal (σ_h), or dihedral (σ_d).
- **Inversion (i):** In this operation, every point in the molecule is moved through a central point (the inversion center) to a position directly opposite to the other side of the molecule. If the molecule remains unchanged, it has a center of symmetry.
- **Improper Rotation** (S_n): This operation is a combination of a rotation around an axis followed by a reflection through a plane perpendicular to that axis. If the

molecule appears unchanged after this combined operation, it has that improper rotation axis as a symmetry element.

These operations define the symmetry elements of a molecule and are used to categorize molecules into point groups, which are important for understanding molecular symmetry and its implications in chemistry and physics. In this case, the work is done within the point group, and the calculations are carried out based on the symmetry elements of the point group.

In summary, the set of symmetry elements of a molecule can be described by the five types of operations:

$$E, \sigma, i, C_n, S_n$$

where

E $\equiv$ The Identity
σ $\equiv$ Reflection.
i $= S_2 \equiv$ Inversion.
C_n $\equiv$ Proper Rotation.
S_n $\equiv$ Improper Rotation.

7.2.1 The Identity E

If the identity operation E is applied to a point with coordinates (x, y, z), its new coordinates (x', y', z') are such that $x = x'$, $y = y'$, $z = z'$.

$$EX = X' \quad con \quad X = X' \tag{7.1}$$

In matrix form:

$$\begin{bmatrix} 1\,0\,0 \\ 0\,1\,0 \\ 0\,0\,1 \end{bmatrix} \begin{bmatrix} x \\ y \\ z \end{bmatrix} = \begin{bmatrix} x' \\ y' \\ z' \end{bmatrix} \tag{7.2}$$

so that

$$E = \begin{bmatrix} 1\,0\,0 \\ 0\,1\,0 \\ 0\,0\,1 \end{bmatrix} \tag{7.3}$$

7.2.1.1 Reflection σ

If the symmetry element of reflection is defined with respect to a plane, then σ_h, σ_v and σ_o,

σ_h ≡ Horizontal Plane.
σ_v ≡ Vertical Plane.
σ_o ≡ Oblique Plane.

The effect of reflecting a point (x, y, z) is to change the sign of the coordinate measured perpendicularly to the respective plane, leaving the coordinates defining the plane invariant.

As an example, consider three principal planes:

$$\sigma(xy) \rightarrow \begin{bmatrix} 1 & 0 & 0 \\ 0 & 1 & 0 \\ 0 & 0 & -1 \end{bmatrix} \begin{bmatrix} x \\ y \\ z \end{bmatrix} = \begin{bmatrix} x \\ y \\ z' \end{bmatrix} \tag{7.4}$$

so that $z = -z'$

$$\sigma(xz) \rightarrow \begin{bmatrix} 1 & 0 & 0 \\ 0 & -1 & 0 \\ 0 & 0 & 1 \end{bmatrix} \begin{bmatrix} x \\ y \\ z \end{bmatrix} = \begin{bmatrix} x \\ y' \\ z \end{bmatrix} \tag{7.5}$$

so that $y = -y'$

$$\sigma(yz) \rightarrow \begin{bmatrix} -1 & 0 & 0 \\ 0 & 1 & 0 \\ 0 & 0 & 1 \end{bmatrix} \begin{bmatrix} x \\ y \\ z \end{bmatrix} = \begin{bmatrix} x' \\ y \\ z \end{bmatrix} \tag{7.6}$$

so that $x = -x'$

In general form:

$$\sigma(x_i x_j) \rightarrow$$

7.2.1.2 Inversion i

Let a point have coordinates (x, y, z). The effect of the symmetry element i is to change the signs of all the coordinates, that is, $i \cdot (x, y, z) = (-x, -y, -z)$.

$$i \rightarrow \begin{bmatrix} -1 & 0 & 0 \\ 0 & -1 & 0 \\ 0 & 0 & -1 \end{bmatrix} \begin{bmatrix} x \\ y \\ z \end{bmatrix} = \begin{bmatrix} x' \\ y' \\ z' \end{bmatrix} \tag{7.7}$$

so that $x = -x'$, $y = -y'$, $z = -z'$

7.2.1.3 Proper Rotation C_n

If we define an axis x_1 that remains invariant under the symmetry element C_n, where C_n is a rotation by θ degrees such that

$$\theta = \frac{2\pi}{n}$$

Example:
Which of the following multiplication tables define groups?

	a	b	c	d
a	b	d	a	c
b	d	c	b	a
c	a	b	c	d
d	c	a	d	b

	a	b	c	d
a	a	b	c	d
b	b	a	d	c
c	c	d	a	a
d	d	c	b	b

7.2.1.4 Rotation Matrix

$$M = \begin{pmatrix} i \cdot i' & i \cdot j' & i \cdot k' \\ j & j & j \\ k & k & k \end{pmatrix} = \begin{pmatrix} cos\ \theta_{i \cdot i} & cos\ \theta_{i \cdot j} & cos\ \theta_{i \cdot k} \\ cos\ \theta_{j \cdot i} & cos\ \theta_{j \cdot j} & cos\ \theta_{j \cdot k} \\ cos\ \theta_{k \cdot i} & cos\ \theta_{k \cdot j} & cos\ \theta_{k \cdot j} \end{pmatrix} \tag{7.8}$$

where the basis $\{i', j', k'\}$ are the unit vectors of the primed system, when a rotation of θ has been performed with reference to the system $\{i, j, k\}$.

As an illustrative example, consider a molecule with triangular symmetry, where the atoms at its corners are identical and can be considered as points, so the mathematical representation is an equilateral triangle. The group's multiplication table is shown below.

$$E = \begin{pmatrix} 1 & 0 \\ 0 & 1 \end{pmatrix} \qquad A = \begin{pmatrix} 1 & 0 \\ 0 & -1 \end{pmatrix}$$

$$B = \begin{pmatrix} -\frac{1}{2} & \frac{\sqrt{3}}{2} \\ \frac{\sqrt{3}}{2} & \frac{1}{2} \end{pmatrix} \qquad C = \begin{pmatrix} -\frac{1}{2} & -\frac{\sqrt{3}}{2} \\ -\frac{\sqrt{3}}{2} & \frac{1}{2} \end{pmatrix}$$

$$D = \begin{pmatrix} -\frac{1}{2} & \frac{\sqrt{3}}{2} \\ -\frac{\sqrt{3}}{2} & -\frac{1}{2} \end{pmatrix} \qquad F = \begin{pmatrix} -\frac{1}{2} & -\frac{\sqrt{3}}{2} \\ \frac{\sqrt{3}}{2} & -\frac{1}{2} \end{pmatrix}$$

Table 7.1 Multiplication table

	E	A	B	C	D	F
E	E	A	B	C	D	F
A	A	E	D	F	B	C
B	B	F	E	D	C	A
C	C	D	F	E	A	B
D	D	C	A	B	F	E
F	F	B	C	A	E	D

A, B, and C are π rotations. D is $\frac{2\pi}{3}$ clockwise rotation. F is $\frac{2\pi}{3}$ counterclockwise rotation. It is observed that AB = D $\neq$ BA (Table 7.1).

Isomorphic $\equiv$ Let two groups G and $\acute{G}$ have a one-to-one correspondence $a \leftrightarrow \acute{a}$ between their elements that preserves the group's multiplication table. That is, if $a \leftrightarrow \acute{a}$ and $b \leftrightarrow \acute{b}$, it implies that $ab \leftrightarrow \acute{ab}$, meaning both groups obey the same multiplication table.

7.3 Rearrangement Theorem

In a multiplication table, each row or column contains each element exactly once, which forms the basis for the Rearrangement Theorem. This theorem states that in a multiplication table of a group, each row or column contains every element of the group exactly once (Fig. 7.1.

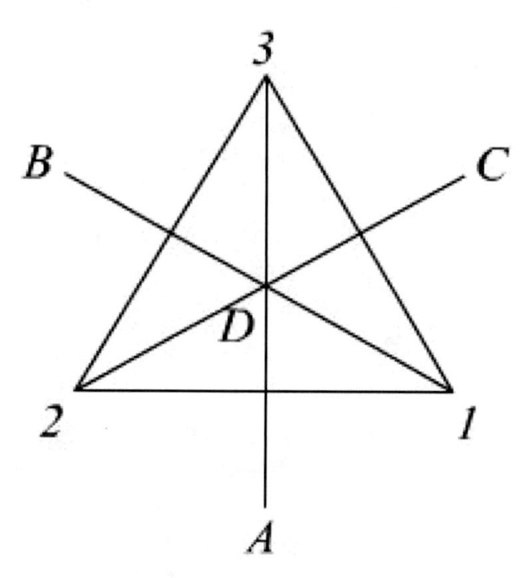

Fig. 7.1 Axes of symmetry

This means that when you arrange the elements of a group in a multiplication table, the result of multiplying each element by a fixed element of the group (i.e., filling in a row or column) will result in a sequence that is simply a rearrangement of the group's elements. No element will appear more than once in any row or column, and all elements will appear exactly once. This property is fundamental in group theory and is a consequence of the group's closure and the existence of inverses.

7.4 Symmetry

In group theory, especially in the context of crystallography and molecular symmetry, symmetry elements are the fundamental components that define the symmetry of an object, particularly in molecular symmetry. Analysis using group theory allows for the reduction of calculations around specific problems.

Here are the main symmetry elements:

Center of Symmetry (Inversion Center): A point within the object such that any line drawn through this point will intersect the object at equidistant points in opposite directions. If the object is inverted through this center, it will appear unchanged. The center of symmetry is one of the most important elements within the set of elements that form the basis of the group. This basis is the set of elements that generate all the properties of the group and are linearly independent.

Plane of Symmetry (Mirror Plane): An imaginary plane that divides the object into two mirror-image halves. Each point on one side of the plane has a corresponding point directly opposite it on the other side.

Axis of Symmetry (Rotation Axis): An imaginary line around which the object can be rotated by a certain angle and still appear the same. For example, a twofold axis means that rotating the object 180 degrees around this axis will result in an indistinguishable configuration.

Rotation Axis: A symmetry element that allows the object to be rotated by an angle (360°/n, where n is the order of the axis) and still appears the same. Common orders include twofold (180°), threefold (120°), fourfold (90°), and sixfold (60°).

Improper Rotation Axis (Rotoinversion Axis): A combination of rotation around an axis followed by reflection through a plane perpendicular to that axis. This is represented by an "S" notation, such as S_{21} (a 21-fold improper rotation).

Mirror Plane: A plane that bisects the object into two halves, which are mirror images of each other, similar to the plane of symmetry.

Rotation Reflection Axis (Rotoreflection): A type of symmetry where the object undergoes rotation followed by reflection in a plane perpendicular to the axis of rotation.

These symmetry elements are used to categorize objects into various point groups in group theory, which helps in understanding their symmetrical properties and behavior under transformations.

Effects of symmetry elements:

1. Rotations around the axis.
2. Reflections in planes.
3. Inversion, which takes $\gamma \quad \rightarrow \quad -\bar{\gamma}$

Inversion $\equiv$ Equivalent to a π rotation followed by a reflection in a plane perpendicular to the rotation axis (Fig. 7.2).

7.5 Notation

Schoenflies

E $\equiv$ Identity.
C_n $\equiv$ Rotation $\frac{2\pi}{n}$ $\quad n = 1, 2, 3, 4, 6.$
σ $\equiv$ Reflection in a plane.
σ_h $\equiv$ Reflection in a horizontal plane, the plane passes through the origin and perpendicular to the axis of high symmetry rotation.
σ_v $\equiv$ Reflection in a vertical plane, the plane passes through the axis of high symmetry.
σ_d $\equiv$ Reflection in a diagonal plane.
S_n $\equiv$ Improper rotation $\frac{2\pi}{n}$.
i $= S_2 \equiv$ Inversion.

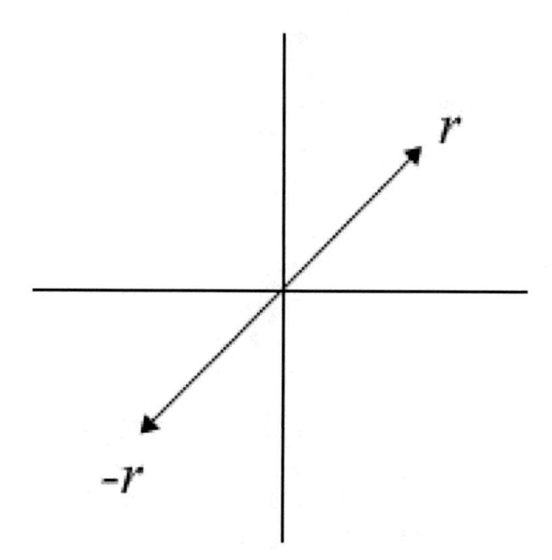

Fig. 7.2 Rotation axes

7.6 Cosets

Let $S = E, A_2, A_3, A_4 ... A_g \qquad S \subset G$
The order of the group G is h.

$Definicion$ Coset $\forall X \in G \wedge X \notin y$

$$\left.\begin{array}{r} SX \equiv coset\ derecho \\ XS \equiv coset\ izquierdo \end{array}\right\} \text{are not subgroups because they do not contain } E.$$
$$SX \cap S = \emptyset$$

Theorem The order g of a subgroup will be an integral divisor of the total group order h, that is, $\frac{h}{g} = l$, where l is an integer called the index of the group y in G.

Example
Let us take the subgroup S from the multiplication Table 7.1. Let the subgroup S = A, E. The left cosets are obtained as follows:

$$SB = \{AB, EB\} = \{D, B\} \quad \rightarrow \quad \text{Not a group because it does not contain } E$$

$$SD = \{AD, ED\} = \{B, D\}$$
$$SB = SD$$
$$SC = \{AC, EC\} = \{F, C\}$$
$$SF = \{AF, EF\} = \{C, F\}$$

It observes that $SX \cap S = \emptyset$

Theorem Let SX and SY be two left cosets, then they either have all elements in common or no elements in common.

$SX = SY$ They have all elements in common.

$SX \neq SY$ They have no elements in common.

Furthermore, the subgroup of order (2) is a divisor of the group of order (6) $\rightarrow$ $\frac{6}{2} = 3$.

Example of finite groups

1. Group of order (1) $\quad G = \{E\}$
2. Group of order (2) $\quad G = \{E, A\}$

	E	A
E	E	A
A	A	E

$A^2 = E$
$A, A^2 = E$ Cyclic order (2)

7.7 Conjugate Elements and Class Structure

It is said that B is conjugate to A if

$$B = XAX^{-1} \qquad \text{or} \qquad A = X^{-1}BX \qquad\qquad \forall X \in G$$

Theorem
If $A \wedge B$ are conjugates of C, then $A \wedge B$ are conjugates.

$$A = XCX^{-1} \quad \wedge \quad B = YCY^{-1} \quad \Rightarrow \quad C = Y^{-1}BY \quad , \quad A = XY^{-1}BYX^{-1}$$

$$XY^{-1} = Z \in G \quad \Rightarrow \quad Z^{-1} = (XY^{-1})^{-1} = YX^{-1} \quad \Rightarrow \quad \boxed{A = ZBZ^{-1}}$$

Class = The set of all mutually conjugate elements.

Example
Let's see that the π degree rotations of the equilateral triangle form a class. Let the element A, determine its mutually conjugate elements. This set of elements forms a class.

1. $D^{-1}AD = FAD = BD = \boxed{C}$
   $D^{-1}D = E = DD^{-1} \qquad \Rightarrow \quad D^{-1} = F$
   $\downarrow \qquad\qquad\qquad \downarrow$
   $F \qquad\qquad\qquad F$
2. $E^{-1}AE = \boxed{A}$
3. $F^{-1}AF = DAF = CF = \boxed{B}$
   $D^{-1} = F \quad \Rightarrow \quad (D^{-1})^{-1} = F^{-1} \quad \Rightarrow \quad F^{-1} = D$
4. $B^{-1}AB \quad \Rightarrow$
   $B^{-1}B = BB^{-1} = E \quad \Rightarrow \quad B = B^{-1}$
   $\downarrow$
   B
   $BAB = FB = \boxed{C}$
5. $C^{-1}AC = CAC = DC = B$
   $C^{-1}C = CC^{-1} = E$
   $\downarrow$
   C
6. $A^{-1}AA = EA = A$

The class C_1 is defined as C_1=A, B, C.

7.8 Conjugate Elements

$$D^{-1}AD = C \quad \Rightarrow \quad D^{-1} = F$$

$D \quad \Rightarrow$ Rotation $\frac{2\pi}{3}$ clockwise.
$F^{-1} \quad \Rightarrow$ Rotation $\frac{2\pi}{3}$ counterclockwise (Fig. 7.3).

7.8.1 Representation Theory

Homomorphism $\forall A_i, A_i' \quad \exists$ correspondence $A_i \quad \leftrightarrow \quad A_i'$ such that if

$$AB = C \quad \Rightarrow \quad A'B' = C'$$

7.8.2 Matrix Representation

$AB = C \quad \Rightarrow \quad \Gamma(A)\Gamma(B) = \Gamma(AB)$

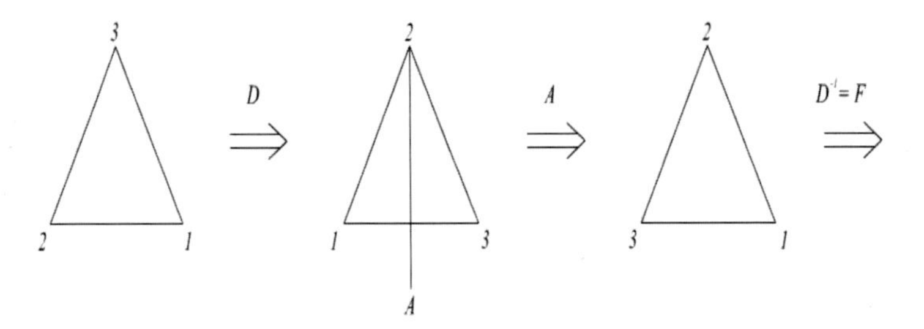

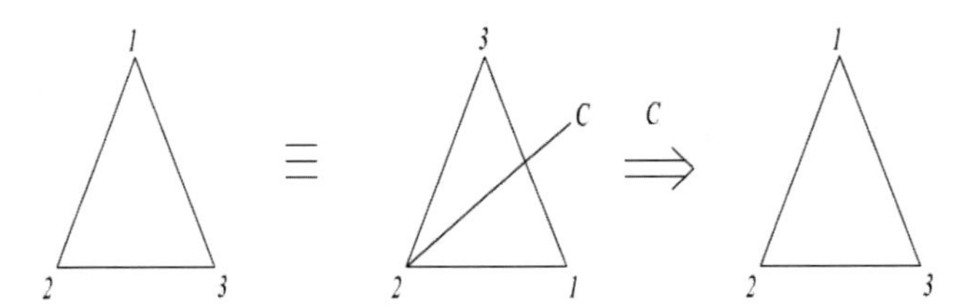

Fig. 7.3 Conjugate elements

7.8.3 *Character of a Representation*

The character of the representation jth is defined as

$$\chi^{(j)}(R) = T_r\Gamma^{(j)}(R)$$

The number of irreducible representations = Number of classes

From the previous exercise with the triangle, there are three classes:

$\ell_1 = E, \quad \ell_2 = A, B, C, \quad \ell_3 = D, F \quad \Rightarrow \quad \exists$ three irreducible representations.

7.8.4 *Dimensionality Theorem*

Let ℓ_i be the dimensionality of an irreducible representation

$$\sum_i \ell_i^2 = h$$

$h \equiv$ Order of the group
$\forall$ For the group of the triangle of order $h = 6$, then,

$$\sum_{i=1}^{3} \ell_i^2 = h \quad \Rightarrow \quad \ell_1^2 + \ell_2^2 + \ell_3^2 = 6$$

Then,

$$2^2 + 1^2 + 1^2 = 6$$

7.8.5 *Construction of the Character Table*

1. The number of irreducible representations is equal to the number of classes.
2. The dimensionality ℓ_i of the irreducible representations is determined by the equation:

$$\sum_{i=1}^{N} \ell_i^2 = h$$

3. For the first column, $\chi^{(i)}(E) = \ell_i$
4. For the first row, $\chi^{(1)}(\ell_k) = 1$
5. The rows are orthogonal and normalized to h with the weight factor N_k (number of elements in the class)

$$\Rightarrow \quad \sum_k \chi^{(i)}(\ell_k)^* \chi^{(j)}(\ell_k) N_k = h\delta_{ij}$$

6. The columns are orthogonal vectors and normalized to $\frac{h}{N_k}$

$$\Rightarrow \quad \sum_i \chi^{(i)}(\ell_k)^* \chi^{(i)}(\ell_l) = \frac{h}{N_k}\delta_{kl}$$

7. $\forall$ For all elements in the i-th row:

$$n_k = \frac{N_k \chi^{(i)}(\ell_k)}{\ell_i}$$

Example Consider the following example of a multiplication table:

	E	A	B	C	D	F
E	E	A	B	C	D	F
A	A	E	D	F	B	C
B	B	F	E	D	C	A
C	C	D	F	E	A	B
D	D	C	A	B	F	E
F	F	B	C	A	E	D

$\Rightarrow$

	ℓ_1	$3\ell_2$	$2\ell_3$
$\Gamma^{(1)}$	1	1	1
$\Gamma^{(2)}$	1	−1	1
$\Gamma^{(3)}$	2	0	−1

The number of classes is $G = \{\ell_1, \ell_2, \ell_3\}$

$$\ell_1 = E, \qquad \ell_2 = \{A, B, C\}, \qquad \ell_3 = \{D, F\}$$

1. $\exists$ There are 3 irreducible representations.
2. $\sum_{i=1}^{3} \ell_i^2 = 6 \qquad \ell_1^2 + \ell_2^2 + \ell_3^2 = 6 \quad \Rightarrow \quad 2^2 + 1^2 + 1^2 = 6$
3. $\forall$ For the first column, $\chi^{(i)}(E) = \ell_i$
4. $\forall$ For the rows, $\chi^{(1)}(\ell_k) = 1$
5. $\forall$ For the rows, $\sum_k \chi^{(i)}(\ell_k)^* \chi^{(j)}(\ell_k) N_k = h\delta_{ij}$

$$\sum_k \chi^{(1)}(\ell_k)^* \chi^{(2)}(\ell_k) N_k = h\delta_{12} = 0$$
$$(1)(1) + (1)(-1)(3) + (1)(1)(2) = 0$$
$$1 - 3 + 2 = 0$$

$$\sum_k \chi^{(1)}(\ell_k)^*\chi^{(3)}(\ell_k)N_k = h\delta_{13} = 0$$
$$(1)(2) + (1)(0)(3) + (1)(-1)(2) = 0$$
$$2 + 0 - 2 = 0$$

6. For the columns, $\sum_i \chi^{(i)}(\ell_k)^*\chi^{(i)}(\ell_l) = \frac{h}{N_k}\delta_{kl}$

$\forall\, k = 1,\ \ l = 2$

$$\sum_i \chi^{(i)}(\ell_1)^*\chi^{(i)}(\ell_2) = \frac{h}{N_1}\delta_{12} = 0$$
$$(1)(1) + (1)(-1) + (2)(0) = 0$$

$\forall\, k = 2,\ \ l = 3$

$$\sum_i \chi^{(i)}(\ell_2)^*\chi^{(i)}(\ell_3) = \frac{h}{N_2}\delta_{23} = 0$$
$$(1)(1) + (-1)(1) + (0)(-1) = 0$$

$\forall\, k = 2,\ \ l = 2$

$$\sum_i \chi^{(i)}(\ell_2)^*\chi^{(i)}(\ell_2) = \frac{h}{N_2}\delta_{22} = \frac{6}{3} = 2$$
$$(1)(1) + (-1)(-1) + (0)(0) = 2$$
$$1 + 1 = 2$$
$$2 = 2$$

7.9 Reducible Representation

7.9.1 Regular Representation

Let's find the regular representation of the following multiplication table:

	E	A	B	C	D	F
E^{-1}	E	A	B	C	D	F
A^{-1}	A	E	D	F	B	C
B^{-1}	B	F	E	D	C	A
C^{-1}	C	D	F	E	A	B
D^{-1}	F	B	C	A	E	D
F^{-1}	D	C	A	B	F	E

$$\Rightarrow \quad \Gamma^{Reg}(A) = \begin{pmatrix} 0&1&0&0&0&0 \\ 1&0&0&0&0&0 \\ 0&0&0&0&0&1 \\ 0&0&0&0&1&0 \\ 0&0&0&1&0&0 \\ 0&0&1&0&0&0 \end{pmatrix}$$

A^{-1}, is found as $\Rightarrow \quad AA^{-1} = A^{-1}A = E \quad \Rightarrow \quad A^{-1} = A$

$B^{-1} = B, \quad C^{-1} = C$

$$D^{-1} \quad \Rightarrow \quad D = D = E \quad \Rightarrow \quad D^{-1} = F$$

$$\Gamma^{Reg}(A) = \begin{pmatrix} 1\,0\,0\,0\,0\,0 \\ 0\,1\,0\,0\,0\,0 \\ 0\,0\,1\,0\,0\,0 \\ 0\,0\,0\,1\,0\,0 \\ 0\,0\,0\,0\,1\,0 \\ 0\,0\,0\,0\,0\,1 \end{pmatrix} \quad \Rightarrow \quad \chi^{Reg}(E) = h$$

7.9.2 Cayley Square Table D_4

See Fig. 7.4.

	E	A	B	C	D	F	G	H
E	E	A	B	C	D	F	G	H
A	A	E	G	H	F	D	B	C
B	B	0	E	0	0	0	0	0
C	C	0	0	E	0	0	0	0
D	D	0	0	0	E	0	0	0
F	F	0	0	0	0	G	H	E
G	G	0	0	0	0	H	E	F
H	H	0	0	0	0	E	F	G

$$\sigma\,(E, A, B, C, D, F, G, H) \tag{7.9}$$

$$A, B, C, D \quad \rightarrow \quad \pi \tag{7.10}$$

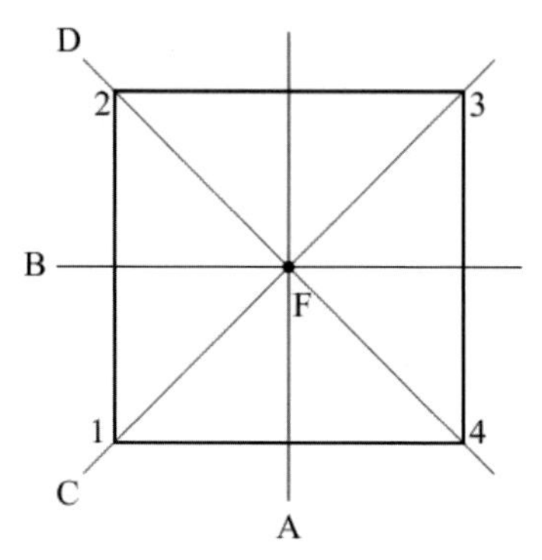

Fig. 7.4 Axes of symmetry of a square

$$F, G, H \quad \rightarrow \quad \frac{\pi}{2}, \pi, \frac{3\pi}{2} \tag{7.11}$$

$$A = \begin{pmatrix} \hat{i}'.\hat{i} & \hat{i}'.\hat{j} \\ \hat{j}'.\hat{i} & \hat{j}'.\hat{j} \end{pmatrix} = \begin{pmatrix} \cos(0) & 0 \\ 0 & \cos(180) \end{pmatrix} = \begin{pmatrix} 1 & 0 \\ 0 & -1 \end{pmatrix} \tag{7.12}$$

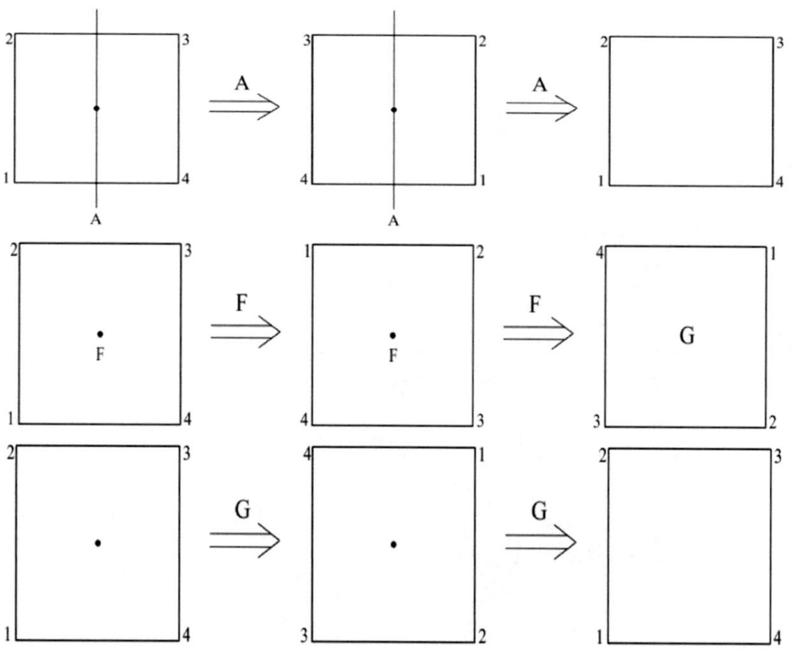

7.9.3 Projection Operators

The use of projection operators constitutes one of the most important tools in the foundations of group theory because it exploits the symmetry of the problem under study. Projection operators take the vectors that form the basis of a representation and project them along the directions that constitute the symmetry of the problem.

Let's look at some properties of the projection operators and their representation in matrix form. Let M be a matrix associated with the eigenvectors $\mathbf{v}_1, \mathbf{v}_2$ *y* $\mathbf{v}_3$ and the eigenvalues λ_1, λ_2 *y* λ_3 related through the eigenvalue equation $(M - \lambda I)\,\mathbf{V} = 0$. Any vector $\mathbf{V}$ in the space generated by the basis V_1, V_2, V_3 can be written as a linear combination, so that

$$\mathbf{V} = \sum_{i=1}^{3} b_i \mathbf{V}_i = b_1\mathbf{V}_1 + b_2\mathbf{V}_2 + b_3\mathbf{V}_3 \tag{7.13}$$

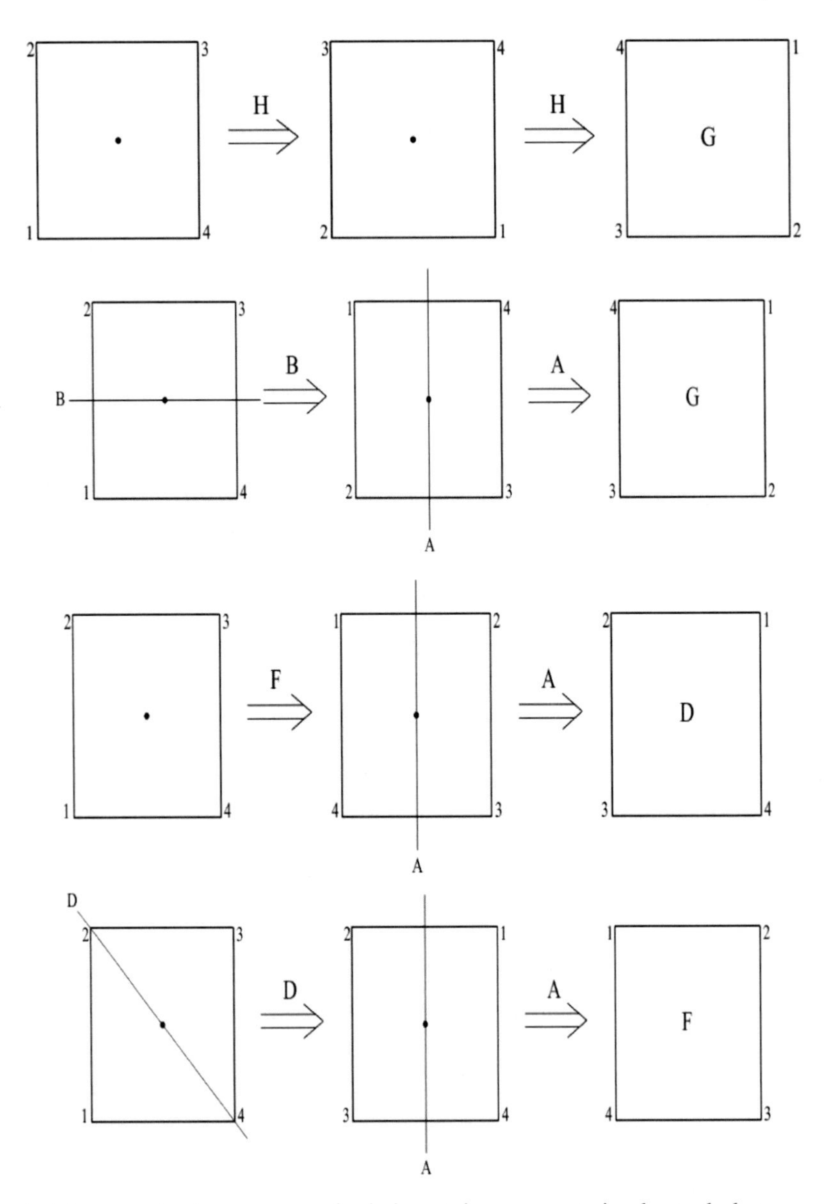

Let three projection operators in their matrix representation be such that

$$P_1\mathbf{V} = b_1\mathbf{V}_1, \quad P_2\mathbf{V} = b_2\mathbf{V}_2, \quad P_3\mathbf{V} = b_3\mathbf{V}_3 \tag{7.14}$$

This implies that the function of the operator P indicates that the matrices P_1, P_2 and P_3 project the vector V in the three directions of the eigenvectors $\mathbf{v}_1$,$\mathbf{v}_2$ and $\mathbf{v}_3$. Now, the eigenvalue equations applied to each of the eigenvalues indicate that

$$(M - \lambda_i I)V_i = 0 \quad \rightarrow MV = \lambda V \tag{7.15}$$

$$(M - \lambda_1 I)V_1 = MV_1 - \lambda_1 V_1 = \lambda_1 V_1 - \lambda_1 V_1 = 0 \tag{7.16}$$

$$(M - \lambda_1 I)V_2 = MV_2 - \lambda_1 V_2 = \lambda_2 V_2 - \lambda_1 V_2 = (\lambda_2 - \lambda_1)V_2 \tag{7.17}$$

$$(M - \lambda_1 I)V_3 = MV_3 - \lambda_1 V_3 = \lambda_3 V_3 - \lambda_1 V_3 = (\lambda_3 - \lambda_1)V_3 \tag{7.18}$$

Expressed in general form:

$$(M - \lambda_j I)V_k = (\lambda_k - \lambda_j)V_k \tag{7.19}$$

So that for $j = k$, we have $(M - \lambda_i I)V_i = 0$. According to the previous properties, we can establish the following relationships:

$$(M - \lambda_j I)(M - \lambda_i I)V_i = 0 \tag{7.20}$$

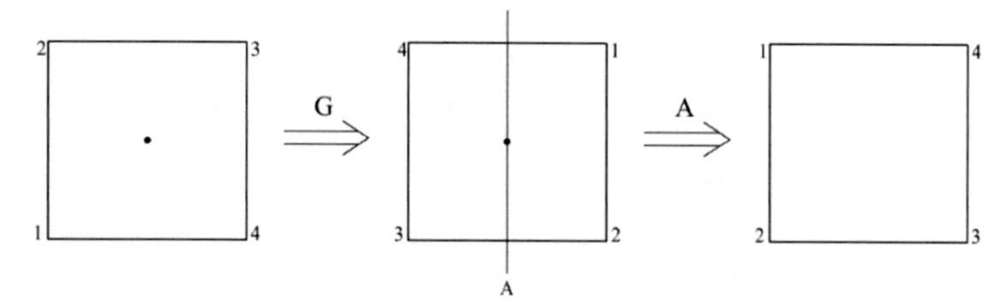

Además

$$(M - \lambda_i I)(M - \lambda_j I)V_k = (M - \lambda_i I)(MV_k - \lambda_i V_k) = \tag{7.21}$$

$$(M - \lambda_i I)(\lambda_k - \lambda_j)V_k = (\lambda_k - \lambda_j)(M - \lambda_i I)V_k \tag{7.22}$$

$$(\lambda_k - \lambda_j)(MV_k - \lambda_i V_k) = (\lambda_k - \lambda_j)(\lambda_k - \lambda_j)V_k \tag{7.23}$$

Additionally,

$$P_1\mathbf{V} = P_1(b_1\mathbf{V}_1 + b_2\mathbf{V}_2 + b_3\mathbf{V}_3) = P_1 b_1\mathbf{V}_1 + P_1 b_2\mathbf{V}_2 + P_1 b_3\mathbf{V}_3 = b_1\mathbf{V}_1 \tag{7.24}$$

$$P_1\mathbf{V}_2 = 0,\ P_1\mathbf{V}_3 = 0 \tag{7.25}$$

From the previous relationship, we can infer that

$$P_i \mathbf{V}_k = \delta_{ik} b_i \mathbf{V}_i \tag{7.26}$$

where δ_{ik} is the Kronecker delta, defined as

$$\delta_{ik} = \begin{vmatrix} 1 \; si \; i = k \\ 0 \; si \; i \neq k \end{vmatrix} \tag{7.27}$$

This demonstrates the orthogonality of the eigenvalues. It is always possible to normalize the eigenvalues such that the b_k are equal to unity. By comparing equations (5.2) and (5.3), we can associate the projection operator matrix P_i with

$$P_i = \frac{(M - \lambda_i I)(M - \lambda_k I)}{(\lambda_i - \lambda_j)(\lambda_i - \lambda_k)} \tag{7.28}$$

The expressions for the P_i are obtained by cyclic permutation of the subscripts. Equation (5.4) assumes that $\lambda_i \neq \lambda_j \neq \lambda_k$, implying that the system is non-degenerate. Another property of the operators is orthogonality, and to this end, let's examine some properties:

$$P_i P_j \mathbf{V}_k = P_i \delta_{jk} b_j \mathbf{V}_j = P_i b_k \mathbf{V}_k = b_k P_i \mathbf{V}_k = b_k \delta_{ik} b_i V_i = b_k b_k \mathbf{V}_k \tag{7.29}$$

In normalization, it is found that $b_k^2 = 1$

$$P_i P_i \mathbf{V} = P_i b_i \mathbf{V}_i = b_i P_i \mathbf{V}_i = P_i \mathbf{V} \tag{7.30}$$

$$P_i P_i = P_i \tag{7.31}$$

Due to this last equation, these operators are known as idempotent. Orthogonality can be observed from

$$P_i P_j \mathbf{V} = P_i b_j \mathbf{V}_j = b_j \delta_{ij} b_i \mathbf{V}_i = \delta_{ij} b_j b_i \mathbf{V}_i = 0, \quad si \; i \neq j \tag{7.32}$$

$$P_i P_j = 0 \quad si \; i \neq j \tag{7.33}$$

To express a definition of the projection operator in terms of the matrix M and its respective eigenvalues, when some of them are identical where the system is called degenerate, it can be done within the system by lifting the degeneracy. The projection matrix associated with two identical eigenvalues $\lambda_r = \lambda_s$ is defined as follows:

$$P_{rs} = \frac{(M - \lambda j I)}{(\lambda_r - \lambda_j)}, \quad r \neq j \ y \ r = s \tag{7.34}$$

So that $P_{rs}\mathbf{V} = b_r\mathbf{V}_r + b_s\mathbf{V}_s$, indicating that the action of the operator P_{rs} on the vector $\mathbf{V}$ is to project it onto the plane defined by the vectors $\mathbf{V}_r$ and $\mathbf{V}_s$, where this plane generated by these vectors is called a two-dimensional degenerate space.

The equations can be generalized in the case where the number of eigenvectors and eigenvalues is greater than three and when there is no degeneracy such that $i = 1, 2, 3$. In such a case, equation (5.4) can be written as

$$Pi = \prod_{\substack{j=1 \\ i \neq j}}^{n} \frac{(M - \lambda j I)}{(\lambda_i - \lambda_j)} \tag{7.35}$$

7.10 Application of Projection Operators

In crystal structures or isolated molecules, the interaction between radiation and matter, as well as the system configuration, is determined by their symmetry. In this context, projection operators are a powerful tool in group theory, used to simplify the analysis of molecular vibrations and quickly obtain the energies associated with the system under study. For a triatomic molecule with equilateral geometry, as demonstrated in the previous example, the use of projection operators can help identify the symmetry-adapted linear combinations (SALCs) of atomic orbitals or molecular vibrations. This method relates to the example used in the chapter on the harmonic oscillator.

7.10.1 *Triatomic Molecule with Equilateral Geometry*

In the case of a triatomic molecule with equilateral triangular geometry, the assigned point group is D_{3h}, and its irreducible representations are labeled as A'_1, A'_2, E', A''_1, A''_2, and E''. These representations describe the symmetry properties of molecular vibrations, electronic states, and other related phenomena. Projection operators can be used to extract components of functions (such as molecular orbitals or vibrations) that transform according to a specific irreducible representation. The projection operator for an irreducible representation γ is given by

$$P^{(\Gamma)} = \frac{l_\Gamma}{|G|} \sum_{g \in G} \chi^{(\Gamma)} R(g) \tag{7.36}$$

where

- l_γ is the dimension of the irreducible representation
- $|G|$ is the order of the group (number of symmetry operations in D_{3h}
- $\chi^{(\Gamma)}$ is the character of the group element g in the irreducible representation γ
- $R(g)$ is the operation corresponding to the group element g.

When applying projection operators, you can follow these four steps:

- (1) identify the base functions
- (2) obtain the character table
- (3) generate the projection operators
- (4) apply these projection operators to the base functions to obtain SALCs (Symmetry-Adapted Linear Combinations) or symmetry-adapted vibrational modes.

For a triatomic molecule, you can choose atomic orbitals (such as s or p orbitals) or displacement coordinates of atoms as the base functions. The character table can be derived from the point group using group theory, with the triatomic molecule typically using the D3h point group. From the previous character table, the projection operators are constructed for each irreducible representation. Finally, the projection operators are applied to the base functions to obtain SALCs or symmetry-adapted vibrational modes.

Example Consider a triatomic molecule with equilateral geometry, with three identical atoms at the vertices of an equilateral triangle, such as BF3. What are the symmetry-adapted linear combinations (SALCs) of the atomic displacements? Use projection operators.

Problem statement Let the basis functions be the Cartesian displacements of the three atoms: For the triatomic molecule BF3, the B ion is located in the center, whereas the F atoms are located at the corners of the triangular system. The BF3 molecule exhibits three polar covalent bonds, and due to its symmetry, the molecule has no net dipole moment. The geometry of this molecule is described as trigonal planar.

(1) Basis Functions The displacements of the three F atoms at the plane triangle's corners are taken as basis functions. That is to say (Table 7.2),

$$f = \{x_1, y_1, x_2, y_2, x_3, y_3\} \tag{7.37}$$

Now, let us construct the projection operators: E' **representation** For the E' representation (which is 2-dimensional, so $l_{E'} = 2$), and the order of the group $|G|$ is 6, therefore, the projection operator is

$$P^{(E')} = \frac{2}{6}\left(E + 2C_3 + 3C'_2 - \sigma_h - 2S_3 - 3\sigma'_v\right) \tag{7.38}$$

Table 7.2 Character table for D_{3h}

D_{3h}	E	$2C_3$	$2C'_2$	σ_h	$2S_3$	$3\sigma'_v$
A'_1	1	1	1	1	1	1
A''_2	1	1	−1	1	1	−1
E'	2	−1	0	2	−1	0
A''_1	1	1	1	−1	−1	−1
A''_2	1	1	−1	−1	−1	1
E''	2	−1	0	−2	1	0

This way, you can apply this operator to each basis function (e.g., x_1, y_1, x_2, y_2, x_3, and y_3) to find the SALCs. For example, by applying this projection operator to the x-displacement vector, it is obtained:

$$P^{(E')}x = \frac{2}{6}\left(Ex + 2C_3x + 3C'_2x' - \sigma_h x - 2S_3x - 3\sigma'_v x\right) \tag{7.39}$$

The proposed exercise recommends that calculations for each term be continued by applying the symmetry operations to the x-displacement vector. After using the projection operator to all basis functions, the SALCs corresponding to the E' representation is obtained. Repeat the process for other irreducible representations when it is considered necessary. In summary, projection operators allow us to systematically derive symmetry-adapted linear combinations (SALCs) for molecular vibrations in a triatomic molecule with equilateral planar geometry. This approach simplifies the analysis of complex molecular systems by leveraging their inherent symmetries. By applying projection operators, we can identify vibrational modes, determine selection rules for spectroscopic transitions, and gain deeper insights into the molecular behavior under various perturbations.

7.11 Proposed Problems

7.1 Repeat the process for other irreducible representations discussed in the previous section for a triatomic molecule with equilateral geometry, such as H_3^+. Use projection operators and apply them to each basis function to find the symmetry-adapted linear combinations (SALCs)

7.2 Consider a rectangle with sides a and b. What are the symmetry elements of this figure? Define the multiplication table and identify the elements for each symmetry class under study.

References

1. Tinkham M (1964) Group theory and quantum mechanics. McGraw-Hill Book Company
2. Cotton FA (1990) Chemical applications of group theory, 3rd ed. Wiley-Interscience

Chapter 8
Symmetries and Quantum Mechanics

The knowledge of a physical system is described by the following eigenvalue equation:

$$\hat{H}\Psi = E\Psi \tag{8.1}$$

$$(\hat{H} - \lambda I)\Psi = 0 \tag{8.2}$$

where the eigenvectors correspond to the wave function Ψ, associated with the state of the system with eigenvalue E_i. In this section, a derivation of equation (6.1) will not be presented, which implies knowledge of the Schrödinger equation where the potential energy function of the system must be specified. Instead, for the eigenvalue equation, it is assumed that the functional expression of the Hamiltonian $\hat{H}$ is known. Thus, knowing the system means determining the set of eigenvectors and eigenvalues of the eigenvalue equation (6.1). Among the properties of the Hamiltonian operator $\hat{H}$, those that are most relevant in this section are related to symmetry operations. For example, when any two atomic motifs are exchanged through a symmetry operation inherent to the system, the Hamiltonian must remain invariant. A symmetry operation is one that leads the system to an equivalent configuration that is indistinguishable from the original configuration from a physical point of view; therefore, the energy of the system must be the same before and after applying the symmetry operation. When system A has greater symmetry than system B, system A has greater degeneracy than B. According to the previous argument, this implies that every symmetry operator XX of the system commutes with the Hamiltonian operator $\hat{H}$, so we can express the following:

$$\left[X, \hat{H}\right] = X\hat{H} - \hat{H}X = 0 \rightarrow X\hat{H} = \hat{H}X \tag{8.3}$$

C. Vargas Hernandez, *Introduction to Raman Spectroscopy and Its Applications*,
https://doi.org/10.1007/978-3-031-77551-2_8

8.1 Symmetry and Quantum Mechanics Applied to Molecules

Symmetry is a powerful tool in quantum mechanics, particularly for studying molecular systems and complex molecular structures. By analyzing the symmetry of molecular systems, calculations that would otherwise be complicated and time-consuming can be simplified. Symmetry also allows for the prediction of physical properties and aids in understanding phenomena observed in optical spectroscopies, such as the splitting of energy levels, selection rules, and degeneracies in molecular orbitals. Through the application of group theory, chemists, physicists, engineers, and other scientists can gain deeper insights into molecular behavior at the quantum level, which would otherwise be more difficult to achieve.

Symmetry plays a crucial role in quantum mechanics, serving as an additional and complementary tool for understanding the energetic behavior of systems, especially in molecular physics. It simplifies the analysis and calculation of molecular structures, their energy levels, and interactions, and helps solve complex quantum mechanical problems that would otherwise be nearly impossible. Here's how symmetry is applied in quantum mechanics for molecules.

8.1.1 Symmetry in Molecular Quantum Mechanics

When studying molecular emissions, we think of a system composed of electrons and nuclei dispersed in space, obeying the laws of physics. This leads to complex equations that interrelate. However, the use of symmetry—specifically how electrons and nuclei are organized to form molecules or more complex systems—becomes a fundamental tool. Symmetry in molecules refers to the invariance of a system under certain transformations, such as rotations, reflections, or translations. When a symmetry operation is applied, the molecule remains as if it were in its initial configuration. These transformations form mathematical structures known as symmetry groups, which are essential for analyzing molecular systems.

- **Point Groups**: A point group is a transformation in which a single point remains unchanged. Molecules can be classified into different symmetry groups based on their geometry, known as point groups (e.g., C_{2v}, D_{4h}, T_d). Each point group corresponds to specific symmetry operations, such as rotation, reflection, and inversion.
- **Molecular Orbitals and Symmetry**: The symmetry of molecular orbitals can be predicted by applying group theory. Molecular orbitals are probability functions that describe the distribution of electron charge, representing the stability of a molecule or molecular system. These orbitals transform according to the irreducible representations of the molecule's point group. For example, in diatomic molecules, orbitals are labeled as σ, π, or δ, depending on their symmetry properties.

8.1.2 Quantum Mechanical Operators and Symmetry

In quantum mechanics, physical observables are represented by operators (such as the Hamiltonian, momentum, or angular momentum). When a system exhibits symmetry, these operators commute with the corresponding symmetry operators, generating conserved quantities that are fundamental in determining outcomes and simplifying calculations. This is similar to classical mechanics, which use energy and momentum conservation laws.

- **Commutators and Conservation Laws**: If the commutator of an operator (such as the Hamiltonian) with a symmetry operator is zero, the system possesses a conserved quantity. For example, in molecules with rotational symmetry, angular momentum is conserved.
- **Degenerate Energy Levels**: Symmetry also explains the degeneracy of energy levels. A level is degenerate when multiple state functions correspond to the same energy. Multiple orbitals or molecular states can share the same energy in molecules with high symmetry, leading to degeneracies. Group theory helps determine degeneracies and how molecular states split under symmetry-breaking perturbations, such as electric or magnetic fields or applied stresses.

8.1.3 Selection Rules in Spectroscopy

Symmetry helps define the selection rules, which govern the allowed transitions between energy levels of molecules, i.e., those transitions that are likely to occur within a molecule when it is appropriately perturbed. These rules are crucial for interpreting spectroscopic data such as infrared (IR) and Raman spectra.

- **Electrical dipolar transitions**: A dipole is formed when a charge is distributed around another charge center. Dipolar transitions are only allowed if the dipole moment operator transforms according to the same irreducible representation as the product of the initial and final states.
- **Vibrational modes**: Symmetry can be used to predict which vibrational modes of a molecule are IR or Raman active. Each vibrational mode transforms according to a specific symmetry representation and depending on the symmetry of the molecule with which it interacts, only certain modes can interact with the incident radiation. When these transitions activate the normal vibrational or rotational modes, they are said to be active in Raman or IR spectroscopy; otherwise, they are said to be inactive.

8.1.4 The Hamiltonian and Symmetry

In quantum mechanics, the Hamiltonian operator describes a system's total energy, including kinetic and potential energies. When the Hamiltonian operator is applied to a wave function, the corresponding energy is obtained by multiplying the wave function, forming an equation known as the eigenvalue equation. If a molecule exhibits certain symmetries, the Hamiltonian can often be simplified.

Reduction of the Hamiltonian: When the Hamiltonian is applied to the wave function of a molecule with specific symmetry, the problem can be reduced into smaller, more manageable problems. This is because the Hamiltonian can be decomposed into smaller blocks, each corresponding to an irreducible representation of the symmetry group. This block-diagonal form simplifies the solution of the Schrödinger equation.

8.1.5 Applications in Molecular Quantum Mechanics

- **Molecular Orbitals**: Group theory and symmetry are essential for constructing molecular orbital diagrams. For instance, in H_2O, symmetry helps determine bonding and antibonding orbitals by examining how atomic orbitals combine based on their symmetry properties. H_2O has three normal modes of vibration, one of which (bending) is applied in thermal heating. To activate this mode, 2.45 GHz radiation is used to increase the temperature of water-containing samples.
- **Molecular Vibrations**: The symmetry of a molecule dictates which vibrational modes are active in IR and Raman spectroscopy. For example, only modes that share the same symmetry as the electric dipole moment will be active in IR spectroscopy.
- **Electronic Structure Calculations**: Symmetry is widely used in materials science and quantum chemistry to reduce computational time and effort. By leveraging molecular symmetry, fewer integrals need to be calculated, thus accelerating molecular property calculations.

8.1.5.1 Examples

Symmetry in the Water Molecule (H_2O)

The water molecule belongs to the $C_{2\nu}$ point group, which has four symmetry elements: identity (E), a twofold rotation axis (C_2), and two symmetry planes ($\sigma\nu$ and $\sigma'\nu$).

- **Molecular Orbitals**: The molecular orbitals of H_2O can be classified according to the irreducible representations of the C_{2v} point group, such as a_1, b_1, and b_2.
- **Vibrational Modes**: Water has three vibrational modes: two symmetric stretching modes (both IR and Raman active) and one bending mode (IR active but Raman inactive). The bending mode is often used to heat samples by applying microwave radiation in the GHz range. Symmetry analysis aids in identifying these modes.

Part III
SERS Spectroscopy Technique

This last part includes chapters on the Raman technique and the SERS spectroscopy technique and an additional chapter with some Raman spectra obtained by the author. The parameters and conditions for the measurements performed to obtain the Raman spectra are shown. Spectra of both inorganic and organic materials and other materials have been obtained, analyzed, and explained. For each of the spectra obtained, a fundamental analysis and the corresponding interpretation of the normal vibration modes and their relation to the structure of the material under study have been carried out.

Chapter 9
Raman Technique

9.1 Introduction

Raman spectroscopy detectors are vital components for capturing the scattered signal, which has a signal-to-noise ratio of approximately 1 in 10^7. This signal is very weak compared to other optical signals. The weak scattered signal captured by the detector results from the interaction between the incident laser light and the molecules present in the materials under study. This scattered signal, known as Raman scattering, occurs when the incident radiation interacts with the normal vibrational modes of the molecules. The detectors convert this scattered light into a measurable voltage or current signal, which is then filtered using optical systems, and its electronic adaptation and amplification are carried out by the respective electronic instrumentation. The captured and processed signal allows for the analysis of the Raman spectrum, with its intensity being proportional to the number of effective interactions that occur between the incident photons and the molecules.

9.2 Definition of Terms in Raman Terminology

The near-infrared region (NIR) covers an area around 12,500 cm^{-1} and extends to 4,000 cm^{-1}. In this region, the bands of C–H, N–H, and O–H are primarily found, making it suitable for the study of many organic compounds. Many applications in organic compounds in the NIR region generally involve functional groups due to the vibrations of H belonging to the respective molecules, such as C–H, S–H, N–H, and O–H.

The mid-infrared region is between 4000 cm^{-1} and 650 cm^{-1} and is further divided into two subregions, one between 4000 and 1300 cm^{-1} and another between 1300 and 650 cm^{-1}. In the first region, the main absorption bands depend on the functional group from which they derive and do not strongly depend on the complete

C. Vargas Hernandez, *Introduction to Raman Spectroscopy and Its Applications*,
https://doi.org/10.1007/978-3-031-77551-2_9

molecular structure. In the second region, it relates to the movement of the bonds that connect the groups to the rest of the molecule. These are deformation and bending vibration frequencies of single bonds in polyatomic molecules. The far-infrared region (FI) covers the area between 700 and 300 cm^{-1}. This region is due to pure rotations and vibration-rotation transitions in gaseous molecules, molecular vibrations in solids and liquids, and lattice vibrations and molecular vibrations in crystals.

9.3 Raman Technique

Raman spectroscopy is used to determine the molecular structure and infer the composition of inorganic and organic materials. It is a versatile and, in principle, non-destructive technique with which solid and liquid materials can be easily examined. Gases can also be examined but under special conditions of the sample holder and density. The minimum sample amount is typically around milligrams and even less if equipment with microscope-type devices and high focusing is available. Raman spectra, which are entirely vibrational spectra, can be acquired with a single instrument. Sample preparation is generally much simpler than in infrared techniques and, in most cases, does not require prior preparation of the samples [1–5].

Currently, there are a variety of Raman spectroscopy variants, such as

1. Non-resonant Raman spectroscopy in visible and near-infrared.
2. Resonant Raman spectroscopy in ultraviolet.
3. Raman imaging microscopy.
4. Time-resolved Raman spectroscopy.
5. High wavenumber Raman spectroscopy.
6. SERS (surface-enhanced Raman spectroscopy).
7. Non-linear Raman spectroscopy.
8. CARS (coherent anti-Stokes Raman spectroscopy).

One of the Raman spectroscopy variants aimed at amplifying the signal-to-noise ratio is Fourier Transform Raman Spectroscopy (FTRS), which also minimizes sample fluorescence, improves frequency precision, and enables low-frequency detection.

9.3.1 FT-Raman Spectroscopy

In conventional Raman spectroscopy, the intensity of the radiation is recorded as a function of the wavenumber. In instruments using FT, light intensity at many wavelengths is captured simultaneously. This process is achieved with CCD camera-type photodetectors, which convert the spectrum into a conventional one via Fourier transform.

9.3.2 Differences and Similarities Between Raman and FTIR Spectroscopy

Both Raman and Fourier transform infrared (FTIR) spectroscopy are vibrational techniques used to study molecular structure, but they differ in some aspects, such as

9.3.2.1 How the Material Interacts with Light

- **In Raman spectroscopy**, the interaction principle is the inelastic scattering of light, where the energy of the incident photon differs from that of the scattered photon. If the energy of the scattered photon is lower, it is called a Stokes-type interaction; if it is higher, it is referred to as an anti-Stokes-type interaction. When coherent radiation, usually from a laser source, hits the sample, part of the energy from the incident photon is used to activate the normal vibrational modes of the molecules that make up the material, causing the emitted photon to have lower energy than the incident photon. The scattered Raman light has approximately a ratio of 1 Raman photon for every 10^7 laser photon. For this reason, it is a technique that requires electronic amplification, noise elimination, high-sensitivity detectors, and appropriate optical filtering. In conclusion, Raman spectroscopy measures the light scattered by the sample.
- **In FTIR spectroscopy**, infrared (IR) radiation is absorbed by the molecules in a sample, meaning it is an optical absorption process. When infrared radiation over a wide spectral range is incident on the sample, the molecules in the material absorb specific frequencies associated with the permitted normal vibration modes. This technique requires a broad-spectrum infrared source and filtering and detection systems using infrared sensors. The method can be applied using reflection or transmission of the signal, depending on the thickness of the sample to be analyzed.

9.3.2.2 How Vibrational Modes Are Activated

- **In Raman spectroscopy**, vibrational modes are activated when a change in the polarizability of the molecule occurs, i.e., the deformation of the electron cloud caused by an external electric field.
- **In FTIR spectroscopy**, vibrational modes are activated when a change in the dipole moment of the molecule occurs, i.e., the separation of positive and negative charges within the molecular structure.

9.3.2.3 Sample Preparation

- **In Raman spectroscopy**, samples can be analyzed without prior preparation, allowing for the study of liquids, solids, and gases. Many liquids are analyzed in aqueous solutions, as the Raman signal is not strongly affected by water.

- **In FTIR spectroscopy**, prior sample preparation is typically required (e.g., using potassium bromide (KBr) pellets for solids). One of the main challenges is that, because it is an absorption process, water can significantly interfere with the spectra due to its strong absorption in the IR region.

9.3.2.4 Wavelength Range Used for the Activation of Normal Vibration Modes

- **Raman spectroscopy** uses light in the visible, near-infrared (NIR), or near-ultraviolet (UV) range, usually around 532, 785, or 1064 nm. Depending on the sample type, the use of lasers with a specific wavelength is recommended to minimize fluorescence interference.
- **FTIR spectroscopy** uses infrared radiation, typically in the mid-infrared region (4000–400 cm^{-1}).

9.3.2.5 The Two Techniques Are Complementary

- **Raman spectroscopy** works best for symmetrical molecules or bonds where the dipole moment does not change (e.g., C=C or S–S bonds).
- **FTIR spectroscopy** is best suited for polar molecules or bonds that exhibit a strong change in dipole moment (e.g., O–H or C=O bonds). This complicates the analysis of aqueous samples, as water is highly polar.

9.3.2.6 Light Response and Sensitivity

- **Raman spectroscopy** is less sensitive to interferences from water and CO_2, making it ideal for studying aqueous solutions.
- **In FTIR spectroscopy**, the strong absorption of water and CO_2 makes it difficult to analyze spectra in aqueous solutions.

9.3.2.7 Invasive Techniques

- **Raman spectroscopy** is generally non-destructive and can be used for in situ analysis, though care must be taken with the laser power, as degradation may occur in some cases, such as with biological samples or certain oxides. To avoid this, it is recommended to reduce the exposure time and take multiple readings.
- **FTIR spectroscopy** is also non-destructive, although certain sample preparation methods may slightly alter the sample (e.g., pressing it into KBr pellets).

9.3.2.8 Applications of the Two Techniques

- **Raman spectroscopy** is used in chemistry, materials science, biology, and pharmaceuticals to study molecular structure, crystal forms, and chemical composition. It is useful in fields such as forensic analysis, art conservation, and nanomaterials. In the latter, a variant known as surface-enhanced Raman spectroscopy (SERS) amplifies the Raman signal using nanoparticles or metal substrates.
- **FTIR spectroscopy** is widely used in organic and inorganic chemistry, pharmaceuticals, polymer science, and biochemistry to identify functional groups, study chemical bonds, and for routine quality control.

In summary, Raman and FTIR are complementary techniques. While Raman is more suitable for nonpolar molecules and samples in aqueous environments, FTIR is more effective at identifying polar molecules and functional groups in various sample types.

9.3.3 Principle of the Raman Technique

The RAMAN system involves illuminating the sample to be studied with a high-energy monochromatic light source. Recently, due to advancements in optoelectronic device technology, lasers of different wavelengths are used. When photons strike the sample, they interact with atomic motifs, i.e., molecules or molecular complexes, and are scattered. The scattering consists of elastic and inelastic collisions. Elastic scattering of photons occurs in all directions, conserving the initial energy of the photons and is known as Rayleigh scattering. RAMAN scattering is a type of inelastic scattering where the energy of the incident photon differs from the scattered photon and occurs when there is a change in the polarizability of the atomic motif. The Raman effect occurs when an intense monochromatic radiation beam passes through a sample containing molecules that can undergo a change in polarizability. In other words, the electron cloud of the molecule must be more easily deformed at the moment of vibration. In the infrared region, vibrations must cause a permanent change in the molecule's dipole moment.

In Raman scattering, the molecule can accept the energy of the incident radiation and excite it to a higher state, emitting a lower energy photon. This radiation is known as Stokes. The energy difference between the incident and scattered photons is necessary to activate one of the molecule's normal vibration modes. The technical difficulty lies in the fact that Rayleigh and Raman scattering are inefficient processes. Under suitable experimental conditions, these processes have an efficiency of 10_3 and 10_7. In other words, out of a million photons incident on the sample, 1000 are elastically scattered photons, and only one photon is inelastically scattered of the Raman type. Due to the inefficiency of the process, high-intensity excitation sources are required. The incident radiation does not send the molecule to any quantized level; instead, it is considered that the molecule is in a virtual state or quasi-state whose height from the initial energy state equals the incident photon's energy. In fact, the wavelength of the incident radiation does not have to be absorbed by the molecule. Through the oscillating induced dipole, energy transfer occurs to the vibrational modes of molecules within the sample. As the electromagnetic wave interacts with the polarized molecule, it ceases to oscillate, and this molecule in this quasi-state returns to its fundamental level, radiating energy in all directions except along the direction of the incident radiation. A quantum of radiation remains in the molecule, so the scattered radiation or photon exits the sample with less energy than the incident one; the quantum necessary to activate a normal vibration mode is determined by measuring the energy difference between the incident and scattered photons. It is also possible that if the scattering molecule is already in an excited vibrational state, a vibrational quantum of energy is transferred in the interaction process with the incident photon, and in the change of polarizability, the molecule emits a photon of higher energy than the incident photon, placing it in a lower vibrational state. This process is called anti-Stokes.

At room temperature, the Stokes process is more efficient than the anti-Stokes process because more molecules are in the ground state than in the excited state. In both processes, the Raman shift is proportional to the normal vibration mode involved in the transition. Thus, the spectrum corresponding to the analysis of a sample consists of a series of lines located energetically at the value corresponding to one of the normal vibration modes and constitutes the material's fingerprint.

9.3.4 Instrumentation Used for Dispersive Raman Technique

Raman spectroscopy consists of a laser that excites the sample and a spectrometer, which typically employs diffraction gratings with 600, 1200, 1800, and 2400 lines per millimeter. The laser radiation enters the spectrometer from the back, designed for this purpose, so that it follows, in some parts of the spectrometer, the same path as the Raman radiation emitted by the sample. The Raman emission is focused toward an entrance slit of a double monochromator and, in some cases, a triple monochromator. The radiation is discriminated according to its wavelength by a stepper motor, and the wavelength selected by the motor is directed toward a detector, which in some cases can be a photomultiplier cooled by the thermoelectric effect to increase the signal-to-noise ratio. However, CCD cameras are commonly used to detect the Raman signal nowadays. It is important to emphasize that it is, in principle, a non-destructive technique, and in the Raman Microscopy version, micrometric-sized particles can be analyzed.

Fourier Transform Infrared Spectroscopy (FTIR) and Raman spectroscopy combined provide a powerful tool for material characterization. Raman is ideal for inorganic species and low-frequency measurements, while FTIR is optimal for analyzing polymeric functional groups.

9.3.5 Light Sources Used in Sample Excitation

The He-Ne laser with a wavelength of 632 nm is generally used because it avoids, to a greater extent, the fluorescence problem for some samples. When different excitation wavelengths are used, qualitative and quantitative analysis can be performed based on the depth of penetration of the laser radiation, as the penetration depth is approximately proportional to the wavelength of the excitation light. This allows for the study of multilayer structures. However, similar studies can currently be conducted using confocal Raman Microscopy. Lasers from the ultraviolet region (244 nm), visible region (632, 532, and 473 nm), to the near-infrared region (785–1064 nm) can be used.

The laser power is important because, of the approximately 10^7 incident photons, only one is a Raman signal. However, depending on the characteristics of the samples, increasing the laser power may increase the fluorescence background, which can overshadow the Raman signal (SEE THE COFFEE SPECTRA).

9.3.6 Detectors

The choice of the Raman signal detection system depends on the configuration of the system itself, including the laser used, as this affects the region of the Raman spectrum where higher resolution is desired.

Common detectors used in Raman spectroscopy vary based on factors such as resolution, bandwidth, and response time. High-performance detectors, which employ more advanced technology, are typically more expensive, with costs depending on the user's needs and the specific analysis required for their samples. Below are some detectors commonly used in Raman spectroscopy:

- **Charge-Coupled Devices (CCDs)**: Commonly used in commercial systems, CCDs are the most commonly used detectors in Raman spectroscopy due to their high sensitivity, good quantum efficiency, and ability to detect weak signals. They are employed in high-resolution techniques and can be coupled with electronic amplification and counting systems for suitable electronic filtering.
  Advantages: CCDs can detect a wide range of wavelengths simultaneously, making them ideal for dispersive Raman spectrometers. Most require external cooling or are internally cooled by current.
- **Photomultiplier Tubes (PMTs)**: Used in specific applications: PMTs are very sensitive to low light levels, making them suitable for detecting very weak Raman signals. Historically, they have been favored for their proportional response to light intensity.
  Advantages: PMTs offer fast response times and are often used in conjunction with monochromators in scanning systems.

- **Electron-Multiplying CCDs (EMCCDs)**: Improved sensitivity: EMCCDs amplify the signal before it is read, making them extremely sensitive to low-light situations.
  Advantages: EMCCDs are particularly useful in applications that require the detection of extremely weak Raman signals, such as in samples with low light or low concentration.
- **InGaAs (Indium Gallium Arsenide) Detectors**: For near-infrared Raman spectroscopy: InGaAs detectors are used for Raman spectroscopy in the near-infrared (NIR) region.
  Advantages: They are sensitive to longer wavelengths, which are not easily detected by standard CCDs.
- **Avalanche Photodiodes (APDs)**: High sensitivity: APDs are used in applications that require very high sensitivity, such as single-photon detection.
  Advantages: APDs are often used in time-correlated single-photon counting (TCSPC) systems for Raman spectroscopy.
- **Back-Illuminated CCDs**: Higher quantum efficiency: These CCDs are designed to increase quantum efficiency by placing the photodetector on the back of the chip, which reduces absorption and reflection losses.
  Advantages: Back-illuminated CCDs offer improved sensitivity, especially in the ultraviolet and visible regions.

The choice of detector depends on several factors, including the spectral range, required sensitivity, and specific application or user needs. For example, in standard dispersive Raman spectroscopy, CCDs are often preferred for their high sensitivity and ability to capture a wide range of wavelengths. Conversely, InGaAs detectors may be more appropriate for applications requiring detection in the near-infrared region. It is important to consider how well the laser, monochromator, and detector assembly are optimized to achieve the best performance.

9.3.7 Monochromator

The monochromator, another of the crucial elements in the resolution of the Raman system, is analyzed below.

A monochromator is an essential component in Raman spectroscopy, used to select and separate specific wavelengths of light, functioning as a wavelength selector. It plays a crucial role in ensuring that only the desired wavelengths reach the detector, which is necessary for accurately measuring the Raman spectrum. A good monochromator facilitates the detector's response by serving as the first noise removal mechanism. Of course, additional filters are necessary to remove the laser signal, which acts as disturbing incident radiation.

What is the function of a monochromator in Raman spectroscopy?
As one of the fundamental elements in spectroscopic techniques, it is important to understand the main functions of a monochromator:

- **Wavelength Selector**: The monochromator is used to select specific wavelengths of the scattered light in optical spectroscopic techniques. In Raman spectroscopy, it isolates the intensity of radiation scattered after interacting with the molecule under study. The monochromator filters out Rayleigh scattered light (which has the same wavelength as the incident light), known as elastic scattering, and allows only the Raman (inelastically scattered) light to pass through. This Raman light, which has a different wavelength due to molecular vibrations, carries information about the inelastic interaction and is directed toward the detector.
- **Resolution Enhancement**: By selecting specific wavelengths, the monochromator improves spectral resolution by reducing optical noise and unwanted signals. This enhances the detection of fine spectral features, which is crucial for distinguishing between closely spaced Raman lines.
- **Stray Light Reduction**: Monochromators help reduce stray light and noise, functioning similarly to a filter. By filtering out unwanted wavelengths, they improve the signal-to-noise ratio, leading to more accurate and reliable data.
- **Customization for Specific Applications**: The type of monochromator used can vary depending on the application and spectral range of interest. For instance, monochromators for near-infrared (NIR) Raman spectroscopy are optimized for longer wavelengths, while those for ultraviolet (UV) Raman spectroscopy are designed for shorter wavelengths.

9.3.8 Some Types of Monochromators Used in Spectroscopy

- **Monochromators with Diffraction Grating**: These are the most commonly used monochromators in Raman spectroscopy. A grating is a series of lines created by holographic methods, and its resolution depends on the number of lines per mm^2. Gratings can be either reflective or transmissive, with reflective gratings being the most commonly used. The grating functions like a prism, spatially separating the light spectrum. These monochromators use a diffraction grating to disperse light into its component wavelengths. The grating can be rotated to direct the desired wavelength toward a collimator or shutter and then to the detector.
  Advantages: High precision in wavelength selection and the ability to cover a wide spectral range.
- **Prism Monochromators**: These are less common and use a prism to disperse light based on the refractive index, which varies with wavelength.
  Advantages: They offer high performance and are often used in applications where a wide spectral range is less critical.
- **Double Monochromators**: These are used when better stray light rejection and very high resolution are required. However, it's important to note that the intensity

of the collected radiation decreases significantly, so the intensity of the incident laser signal on the sample may need to be increased. Double monochromators consist of two monochromators in series, providing even better stray light rejection and improved signal purity.
Advantages: They are particularly useful in applications requiring extremely low noise levels, such as detecting very weak Raman signals, analyzing very small samples, or searching for trace amounts of a sample in a matrix.

Monochromators are vital for the accurate measurement of Raman spectra, allowing for the precise identification of molecular vibrations and chemical compositions. Their ability to filter and separate specific wavelengths ensures that Raman signals are detected with clarity, leading to better analysis and quantification accuracy.

References

1. Cardona M (1982) Light scattering in solids II: basic concepts and instrumentation. Springer
2. Bhargava R, Petruska MA (2013) Raman imaging: techniques and applications. McGraw-Hill
3. Toporski J, Dieing T, Hollricher O (2018) Confocal Raman microscopy. Springer
4. Lewis IR, Edwards HGM (2001) Handbook of Raman spectroscopy: from the Research Laboratory to the process line. Marcel Dekker
5. Turrell G, Corset J (1996) Raman microscopy: developments and applications. Academic Press

Chapter 10
SERS Spectroscopy Technique

Surface-Enhanced Raman Scattering (SERS) is a powerful spectroscopic technique that enhances the Raman scattering of molecules adsorbed on rough metal surfaces or nanostructures [1, 2]. Here's a detailed overview of the SERS technique.

10.1 Principle of SERS

Raman scattering occurs when light interacts with molecular vibrations, leading to inelastic scattering, where the energy of the scattered light differs from that of the incident light. The energy difference corresponds to the molecule's vibrational modes, providing a molecular fingerprint.

Enhancement Mechanism SERS enhances the Raman scattering signal by factors up to 10^6 to 10^{10} or even higher. This enhancement arises from two main mechanisms:

(1) **Electromagnetic Enhancement**:

Light irradiating a rough metal surface or metal nanoparticles induces localized surface plasmon resonances (LSPRs). These are coherent oscillations of the metal's conduction electrons, which create intense electromagnetic fields at the surface.

The electromagnetic fields enhance the Raman scattering of molecules near the metal surface.

C. Vargas Hernandez, *Introduction to Raman Spectroscopy and Its Applications*,
https://doi.org/10.1007/978-3-031-77551-2_10

(2) **Chemical Enhancement**:
This involves charge transfer between the adsorbed molecule and the metal surface. The interaction can modify the molecule's electronic properties, leading to increased Raman scattering cross sections.

10.2 SERS Substrates

Metallic Nanostructures:

Common metals used for SERS substrates include gold (Au); silver (Ag); and, to a lesser extent, copper (Cu). These metals are chosen due to their strong plasmonic properties in the visible and near-infrared regions of the electromagnetic spectrum.

Types of SERS Substrates

Nanoparticles: Colloidal solutions of metal nanoparticles (e.g., gold or silver nanoparticles).

Nanostructured Surfaces: Surfaces with engineered nanostructures such as nanopillars, nanoholes, or roughened films.

Nanocomposites: Combinations of metal nanoparticles with other materials (e.g., graphene, silicon) to enhance stability and functionality.

10.3 Applications of SERS

Chemical and Biological Sensing

SERS is widely used for detecting low concentrations of chemical and biological molecules, making it valuable in environmental monitoring, food safety, and medical diagnostics [3].

Single-Molecule Detection The high sensitivity of SERS enables the detection of single molecules, providing insights into molecular interactions and dynamics at the single-molecule level.

Material Science SERS is used to study surface phenomena, including adsorption, catalysis, and corrosion processes.

Forensic Science SERS assists in the detection and identification of trace amounts of forensic samples such as drugs, explosives, and inks.

10.3.1 *Advantages of SERS*

High Sensitivity: Capable of detecting low-abundance molecules.
Specificity: Provides molecular-specific information.
Non-Destructive: Preserves the sample integrity.

10.3.2 *Challenges and Future Directions*

Reproducibility: Achieving consistent and reproducible SERS signals remains a challenge due to the variability in substrate fabrication.
Quantification: Developing reliable quantification methods for SERS signals.
Substrate Development: Advancing the design and fabrication of robust, uniform, and reproducible SERS substrates.

SERS continues to be an area of active research with ongoing developments aimed at improving its applications and overcoming existing challenges.

10.4 Equations for SERS

Surface-Enhanced Raman Scattering (SERS) involves several key equations related to the enhancement mechanisms. Here's an overview of the primary equations involved in SERS:

Raman Scattering Intensity

The intensity of the Raman scattering I_R is given by

$$I_R = I_0 N \sigma_R \tag{10.1}$$

where

- I_0 is the intensity of the incident light,
- N is the number of scattering molecules, and
- σ_R is the Raman scattering cross section.

10.4.1 *Electromagnetic Enhancement Factor*

The electromagnetic enhancement factor G_{EM} arises from the enhancement of the electric field near the metal surface due to localized surface plasmon resonances. This can be approximated by

$$G_{EM} \propto |E_{Loc}|^4 \tag{10.2}$$

where

- E_{Loc} is the local electric field at the molecule's location and
- $|E_{Loc}|$ is typically much larger than the incident field E_0, particularly at plasmon resonance.

Local Electric Field Enhancement

The local electric field enhancement E_{Loc} near a metallic nanostructure can be described by

$$E_{Loc} \approx E_0 \left(\frac{\varepsilon_m - \varepsilon_d}{\varepsilon_m + \varepsilon_d} \right) \tag{10.3}$$

where

- E_0 is the incident electric field,
- ε_m is the dielectric constant of the metal, and
- ε_d is the dielectric constant of the surrounding medium.

Total Enhancement Factor The total enhancement factor G_{EMT} combines both electromagnetic and chemical enhancements. For simplicity, we often consider the electromagnetic part, which is the dominant contribution:

$$G_{EMT} = \left(\frac{E_{Loc}}{E_0} \right)^4 \tag{10.4}$$

10.4.2 SERS Intensity

Considering the enhancement factors, the SERS intensity I_{SERS} can be written as

$$I_{SERS} = I_0 N \sigma_R G_{EMT} \tag{10.5}$$

where

- G_{EMT} includes both electromagnetic and chemical enhancement factors, though the electromagnetic enhancement is usually the primary contributor.

Chemical Enhancement Factor The chemical enhancement factor G_{Chem} is more complex and depends on the interaction between the molecule and the metal surface. It can involve factors such as charge transfer and electronic coupling, but a simple approximation is often challenging to derive without detailed knowledge of the system.

Combined Enhancement When considering both electromagnetic and chemical enhancements, the total SERS enhancement factor G_{Total} is given by

$$G_{total} = G_{EM} G_{Chem} \tag{10.6}$$

Therefore, the SERS intensity considering both mechanisms is

$$I_{SERS} = I_0 N \sigma_R G_{Total} \tag{10.7}$$

In practical applications, G_{EM} often dominates, and total G_{Total} is primarily influenced by the electromagnetic enhancement. These equations provide a theoretical foundation for understanding and calculating the enhancement observed in SERS experiments.

Example To calculate the enhancement factorfor Surface-Enhanced Raman Scattering (SERS), we need to consider the electromagnetic (EM) enhancement mechanism, which involves the enhancement of the local electric field near the metal nanoparticles. The key parameter is the ratio of the local electric field E_{Loc} to the incident electric field E_0.

Given the dielectric constant of the nanoparticle, we can use the following formula to estimate the local electric field enhancement:

$$\left|\frac{E_{Loc}}{E_0}\right| \approx \left|\frac{\varepsilon_m - \varepsilon_d}{\varepsilon_m + \varepsilon_d}\right| \tag{10.8}$$

where ε_m is the dielectric constant of the metal nanoparticle of silver and ε_d is the dielectric constant of the surrounding medium which, in our case, is water.

According to the theoretical model, the dielectric constant of the silver at 632.8 nm is $\varepsilon_m = -16.32 + 0.54i$, this complex dielectric constant is crucial for modeling and understanding the plasmonic properties and electromagnetic enhancement in SERS applications. The dielectric constant of liquid water is around 78.4 (Fernandez et al. 1995, 1997).

Let us calculate the SERS enhancement factor for silver nanoparticles at a wavelength of 632.8 nm, using the previously provided dielectric constant values:

Calculate the Electric Field Enhancement Factor:

Let the equation,

$$\left|\frac{E_{Loc}}{E_0}\right| \approx \left|\frac{\varepsilon_m - \varepsilon_d}{\varepsilon_m + \varepsilon_d}\right| \tag{10.9}$$

By substituting previous values, we obtain

$$\left|\frac{\varepsilon_m - \varepsilon_d}{\varepsilon_m + \varepsilon_d}\right| = \left|\frac{-16.32 + 0.54i - 78.4}{-16.32 + 0.54i + 78.4}\right| = \left|\frac{-94.72 + 0.54i}{62.08 + 0.54i}\right|$$

$$\left|\frac{\varepsilon_m - \varepsilon_d}{\varepsilon_m + \varepsilon_d}\right| = \left|\frac{(-94.72 + 0.54i)}{(62.08 + 0.54i)}\frac{(62.08 - 0.54i)}{(62.08 - 0.54i)}\right| = \left|\frac{-5.88x10^3 + 84.67i}{3.854x10^3}\right|$$

$$\left|\frac{\varepsilon_m - \varepsilon_d}{\varepsilon_m + \varepsilon_d}\right| = \sqrt{\left(\frac{-5.88x10^3}{3.854x10^3}\right)^2 + \left(\frac{84.67}{3.854x10^3}\right)^2} = 1.526 \tag{10.10}$$

Calculate the Electromagnetic Enhancement Factor:

Using previous values and using the definition of the improvement factor, we have

$$G_{EMT} = \left(\frac{E_{Loc}}{E_0}\right)^4 = \left|\frac{E_{Loc}}{E_0}\right|^4 \approx \left|\frac{\varepsilon_m - \varepsilon_d}{\varepsilon_m + \varepsilon_d}\right|^4 \tag{10.11}$$

Therefore,

$$G_{EMT} = (1.526)^4 = 5.423 \tag{10.12}$$

So, the electromagnetic enhancement factor for SERS at a wavelength of 632.8 nm for silver nanoparticles, given the dielectric constant values, is approximately 5.423. This result of the calculated enhancement factor demonstrates the significant increase in the local electric field due to the plasmonic properties of silver nanoparticles. In actual SERS applications, the total enhancement factor might be even higher due to additional contributions from chemical enhancement and the specific morphology of the nanoparticles.

Calculate the intensity I_{Laser} of the incident laser radiation:

To calculate the intensity of the incident radiation of the laser used in a Surface-Enhanced Raman Scattering (SERS) experiment, we need to consider the power of the laser and the area over which the laser beam is focused. The intensity I_{Laser} of the incident laser radiation is given by

$$I_{Laser} = \frac{P_{Laser}}{A_{spot}} \tag{10.13}$$

where I_{Laser} and A_{spot} are the incident radiation of the laser ($\frac{W}{m^2}$) and the area over which the laser beam is focused (in square meters, m^2), respectively.

To calculate the Raman scattering cross section, we need to understand the relation between the Raman intensity and the properties of the material. The Raman scattering cross section, σ_R, is a measure of the probability of Raman scattering per molecule per unit incident power.

To find the Raman scattering cross section, we can rearrange Eq. 10.1, where $I_R = 1 \times 10^{-18} W$ Raman intensity without enhancement data obtained from an experiment technique.

$$\sigma_R = \frac{I_R}{I_0 N} \tag{10.14}$$

In this case, I_0 is the intensity of the incident laser radiation.

Below is the calculation for the effective SERS enhancement when the incident laser has a power of $I_{Laser} = 10\text{mW}$, a wavelength of $\lambda = 632.8$ nm, and a laser spot diameter D_{Spot} of $1\,\mu$m.

By using (10.14), and

$$\left| \begin{array}{l} I_0 = I_{Laser} \\ I_{Laser} = \frac{P_{Laser}}{A_{spot}} \\ A_{spot} = \pi\left(\frac{D_{Spot}}{2}\right)^2, \quad \text{SpotArea} \end{array} \right. \tag{10.15}$$

then

$$\left| \begin{array}{l} \sigma_R = \frac{I_R}{I_{Laser} N} = \frac{I_R A_{spot}}{P_{Laser} N} = \frac{I_R A_{spot} \pi D_{Spot}{}^2}{4 P_{Laser} N} \\ \\ \sigma_R = 7.854 \times 10^{31}\ \text{cm}^2 \end{array} \right. \tag{10.16}$$

References

1. Kneipp K, Moskovits M, Kneipp H (eds) (2006) Surface-Enhanced Raman scattering: physics and applications. Springer
2. Le Ru EC, Etchegoin PG (2009) Principles of surface-enhanced Raman spectroscopy and related plasmonic effects. Elsevier
3. Van Duyne RP, Aroca R, Schatz GC (2002) Surface-Enhanced Raman Scattering (SERS): future research directions and applications. J Phys Chem B 106(37):9279–9293

Chapter 11
Raman Spectra of Some Materials

In this chapter, several Raman spectra measured by the author will be presented, along with the corresponding peak assignments based on specialized literature. A detailed study of the symmetries or an in-depth discussion of the Raman spectra is not included. It is recommended to consult specialized literature for a more comprehensive analysis of the spectra of the materials discussed here. The objective is to demonstrate the versatility of Raman spectroscopy and its capability to characterize a wide variety of materials.

The samples measured and analyzed include organic and inorganic materials in both liquid and solid states. Some samples require prior preparation, while most do not need any conditioning prior to analysis. Important factors in the measurements include the laser power, which is regulated by a set of neutral density filters for the specific laser employed. In this case, the laser had a wavelength of 473 nm, and the system used included a 1-m monochromator, a current-cooled CCD detector camera, and a confocal microscope for sample alignment and precise localization of the laser impact area. The system offers 1 μm lateral resolution for region and target selection ranging from $10\times$ to $100\times$. The Raman system allows the selection of sampling time, wavelength range, and number of repetitions. Sampling time is a crucial parameter; longer sampling times increase signal intensity but can also broaden spectral peaks due to heating of the sample by laser exposure. This heating can sometimes degrade inorganic oxide samples during the measurement process.

C. Vargas Hernandez, *Introduction to Raman Spectroscopy and Its Applications*,
https://doi.org/10.1007/978-3-031-77551-2_11

11.1 Inorganic Materials

11.1.1 Silicon Calibration Standard

Monocrystalline silicon is usually used to calibrate Raman systems because it provides a peak around 520.7 cm^{-1} (Fig. 11.1).

Silicon is one of the most widely used materials in device technology due to its versatility in manufacturing doped semiconductors for applications such as rectifier diodes, emitter diodes, and a wide array of other devices, including the production of transistors. Another significant application of silicon-based junctions is in the manufacture of solar cells, which are crucial for modern energy generation. Additionally, silicon is the second most abundant material in the Earth's crust. Given its well-defined Raman peak and its importance in the semiconductor industry, silicon has been extensively studied in Raman spectroscopy. The Raman spectrum of silicon typically displays a prominent peak that provides crucial information about the material's structural and electronic properties. In our case, silicon serves as the material for calibrating our Raman system.

Silicon exhibits a strong and sharp Raman peak at around 520.7 cm^{-1}, corresponding to the optical phonon mode in its crystal structure. The exact position and intensity of this peak can reveal details about the silicon sample's crystal quality, stress, and temperature. In the semiconductor industry, Raman spectroscopy is widely employed to monitor the quality of silicon wafers during manufacturing. The position, width, and intensity of the Raman peak can indicate the presence of stress, defects, or impurities in the silicon crystal. For instance, stress-strain analysis can be inferred from shifts in the 520.7 cm^{-1} peak, where compressive deformation typically causes a shift to higher wavenumbers, and tensile deformation shifts the peak to

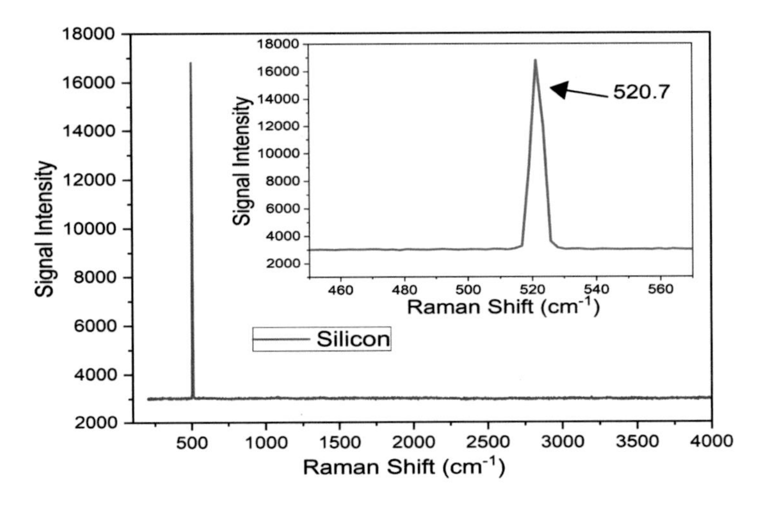

Fig. 11.1 Silicon

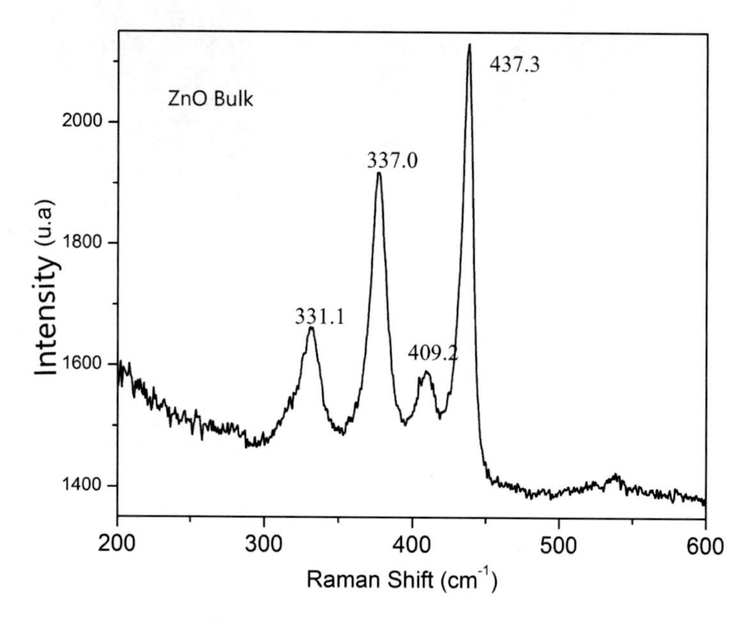

Fig. 11.2 Zinc Oxide

lower wavenumbers. Another important characteristic is the temperature-dependent behavior of semiconductors. As the temperature increases, the Raman peak usually broadens and shifts toward lower wavenumbers, making this property useful in the non-contact thermal analysis of silicon-based devices. In the microelectronics industry, Raman spectroscopy is applied at various stages of device fabrication to analyze thin films, detect residual stress, and ensure the integrity of silicon during processing.

For the study of silicon, lasers with wavelengths that minimize fluorescence and maximize Raman scattering efficiency are commonly used. Typical choices include 514.5 nm (green) or 785 nm (near-infrared); in our case, we used a 473 nm laser. Silicon samples generally do not require special preparation for Raman analysis, making the technique convenient and non-destructive. However, the surface must be clean and free of dust, grease, and other contaminants, which can be easily removed with solvent cleaning (Fig. 11.2).

11.1.2 Zinc Oxide

11.1.2.1 Normal Vibrational Modes of ZnO

Zinc oxide is one of the most versatile and technologically important semiconducting materials [1]. One of its most significant properties for optoelectronic applications is its bandgap, which typically ranges around 3.14 eV and strongly depends on the

Table 11.1 Normal vibrational modes of ZnO. The Raman peaks are consistent with the references [3, 7–12]

Normal mode of vibration	$\lambda[\mathrm{cm}^{-1}]$	Associated
E2(low)	100	Low non-polar mode
E2M	332–334	Second-order mode multi-phonon processes
A1(TO)	378-383	Second-order transverse mode with A1 symmetry
E1(TO)	407–410	First-order transverse mode
E2(high)	437	Non-polar optical phonon
E1(LO)	580–584	Longitudinal mode with E1 symmetry
A1(LO)	550	Longitudinal mode with A1 symmetry. Surface phonon mode
2TO	983	Second-order mode
A1, E1 (acous. Comb.)	1100	Second-order mode that corresponds to an acoustic combination of the A1(LO) and E2(low) modes
A1(2LO)	1154	Second-order mode associated with a longitudinal mode

growth method [2–4]. ZnO is a semiconductor material with a hexagonal wurtzite structure, consisting of four atoms per unit cell. Group theory predicts that for this configuration, the space group symmetry is C4(6v)(P63mc), indicating that there are theoretically optical phonons $A1$, $E1$, $2E2$, $2B1$ [5, 6]. The symmetric phonon modes $A1$ and $E1$ are both Raman and infrared active. The $E2$ mode is only Raman active, and the $B1$ mode is silent (forbidden for both Raman and infrared). The polar characteristics of the vibrational modes $A1$ and $E1$ lead to a splitting into longitudinal optical (LO) and transverse optical (TO) modes. Considering the above, several possible modes for ZnO have been reported in the literature and are summarized in Table 11.1.

Figure 11.3 shows the Raman spectroscopy spectra of ZnO films deposited on glass substrates with both chemical and physical pre-treatments. The samples were deposited using the SILAR technique [12] and underwent thermal treatments at temperatures of 100, 150, and 200 ^0C. The figure shows modes at 272, 332, 380, 410, 437, 510, 550, 580, 642, 980, 1100, 1154 cm^{-1}. The modes at 332 and 380 cm^{-1} are due to multi-phonon processes indicating the better crystalline quality of the samples; these modes become evident at temperatures above 150 ^0C. The peak at 550 cm^{-1} has been assigned to a surface phonon mode and gradually disappears with thermal treatment. This is because the increase in thermal treatment temperature decreases the grain size, reducing the density of grain boundaries and thus the charge trapping

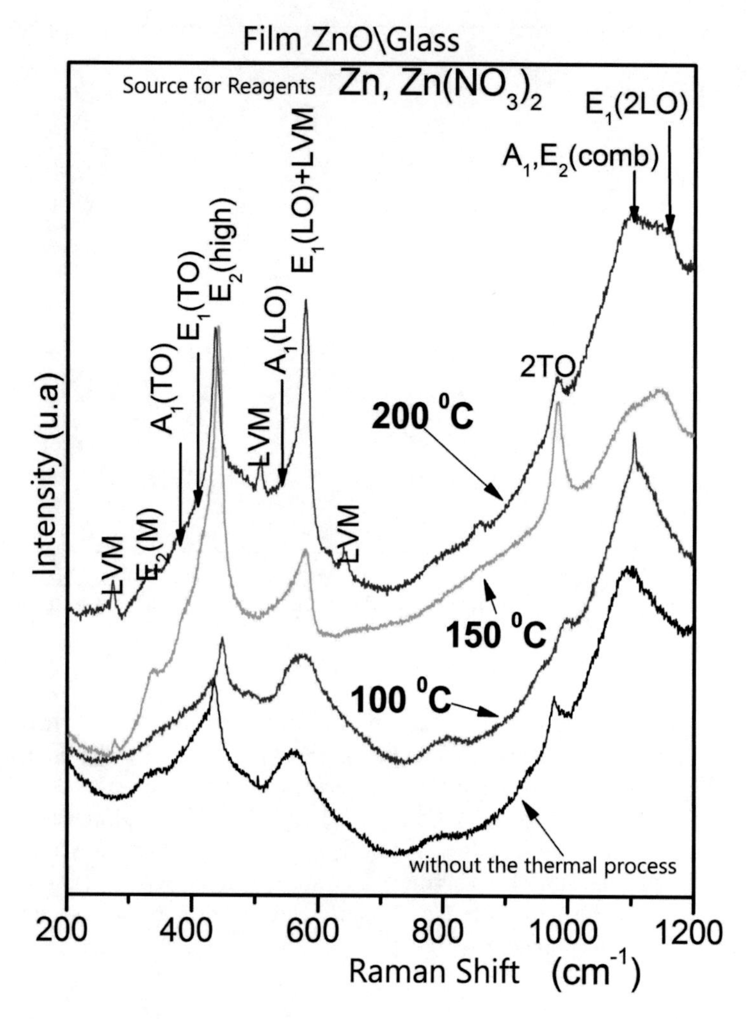

Fig. 11.3 Raman spectra of ZnO film deposited on a glass substrate, without thermal treatment, and at 100, 150, and 200 °C

effects in these boundaries, in addition to causing surface reordering as diffusion of the species is favored.

The interpretation of the normal vibrational mode located around 580 cm^{-1} still needs to be clarified. It has been reported in the literature due to an $E_1(LO)$ mode related to oxygen vacancies and a local vibrational mode (LVM) of nitrogen. Around 1000 and 1200 cm^{-1}, two modes A1.E2(acoust. Comb) and E1(2LO) are observed. The appearance of nitrogen in the samples, which becomes evident at temperatures above 150 °C, is due to the precursor source used (NH_4OH), which is introduced as an impurity into the ZnO crystalline structure at high treatment temperatures; when the hydroxides anchored in the initial growth stage transform into ZnO.

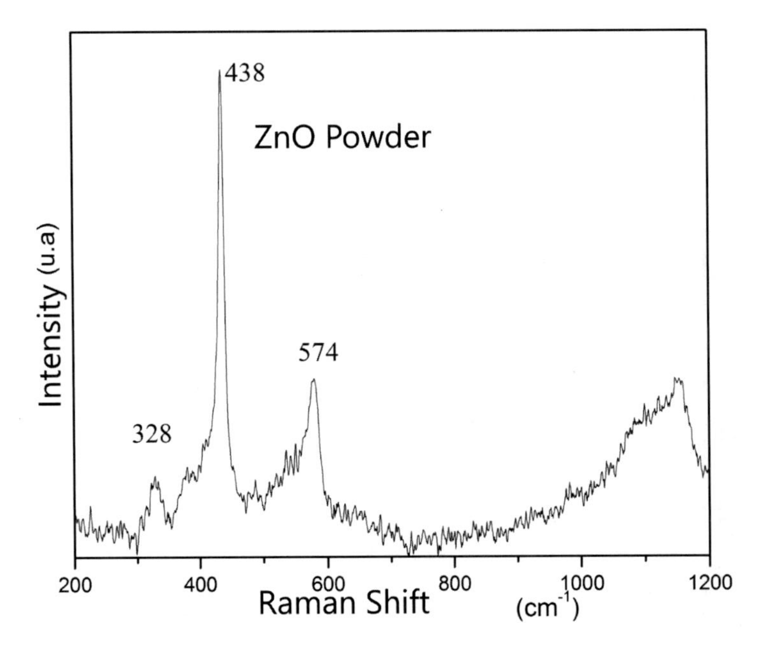

Fig. 11.4 Raman spectrum of ZnO powder

Figure 11.4 shows the Raman spectrum of a ZnO powder sample, where three characteristic peaks are observed around 328, 438, and 574 cm^{-1}. These peaks are associated with unidentified impurities in a second-order transverse mode with A_1 symmetry, first-order transverse mode $E_1(TO)$, and a longitudinal mode $E_1(LO)$ related to oxygen vacancies. As observed, depending on whether the sample is in powder or film form and the synthesis routes, the normal vibrational modes associated with an ideal ZnO crystalline structure manifest or not due to the appearance of impurities or symmetry breaking (Fig. 11.4).

11.1.3 Vanadium Oxide

Vanadium pentoxide (V_2O_5) and its hydrated forms (V_2O_5 nH_2O) crystallize in a laminar structure and have technological applications across a wide range of uses, such as catalysts in chemical reactions [13], cathodes for solid-state batteries [14], extended gate field-effect transistors (EGFETs) [15], electrochromic devices [16], electronic and optical switches, among others. When vanadium oxide in the form of amorphous films is exposed to different atmospheres (methanol, ethanol, acetone, and isopropanol), its physical properties change. The Raman spectra at room temperature for VO are shown in Fig. 11.5. The normal vibration modes are located at 145 cm^{-1} (V–O–V...), 284 cm^{-1} (V=O), 406 cm^{-1} (V=O), 701 cm^{-1} (V–O–V), and 991 cm^{-1} (V=O) [17–19, 47].

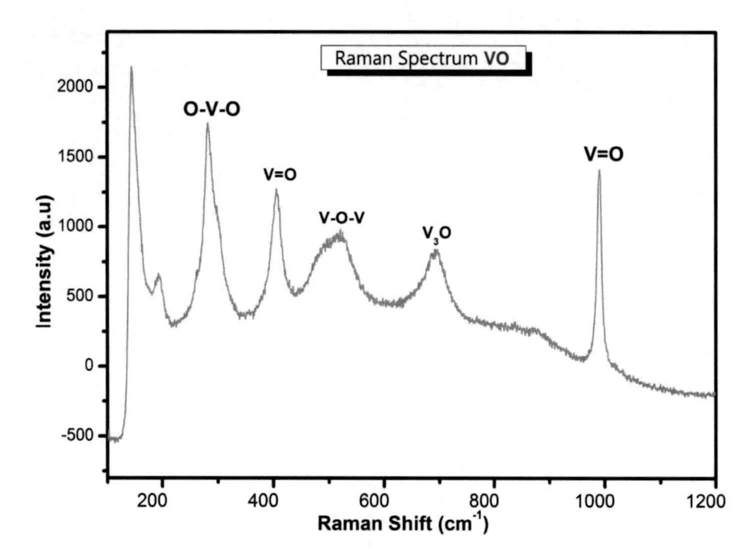

Fig. 11.5 Vanadium Oxide

11.1.4 Zinc Chromite $ZnCr_2O_4$

Spinels are face-centered cubic structures with the general chemical formula AB_2X_4, where A can be Zn, Cd, Fe, Cu, or Ge; B can be Cr, Al, Co, Ga, or In; and X can be O, S, Se, or Te. These structures consist of a compact arrangement of X ions, with interstices occupied by A or B cations, forming a cubic cell. Zinc chromium oxide ($ZnCr_2O_4$) belongs to this family and has a face-centered cubic structure, specifically the O_7 h (Fd3m) space group, where Zn atoms occupy tetrahedral sites and Cr atoms occupy octahedral sites. This configuration is known as a normal-type spinel. However, when the cationic distribution is inverted, an inverse spinel configuration is obtained. Due to vacancies, the crystal field is modified, allowing Cr atoms to occupy tetrahedral sites. This results in a mixture of configurations between normal and inverse spinel, which can exist at low temperatures. Since the 1930s, ACr_2X_4 chromite spinel structures have been studied with great interest [11, 20, 21].

Figure 11.6 shows the Raman spectrum with characteristic bands corresponding to the normal vibration modes of Zn and Cr atoms in tetrahedral and octahedral environments formed by oxygen atoms, located around 400 and 900 cm^{-1}, respectively. The bands around 941 cm^{-1} are associated with vacancies in tetrahedral and octahedral sites due to the interaction between the Zn and Cr atoms and may play an important role in surface gas absorption [22]. These oxygen vacancies are generated because the sintering process is carried out at high temperatures.

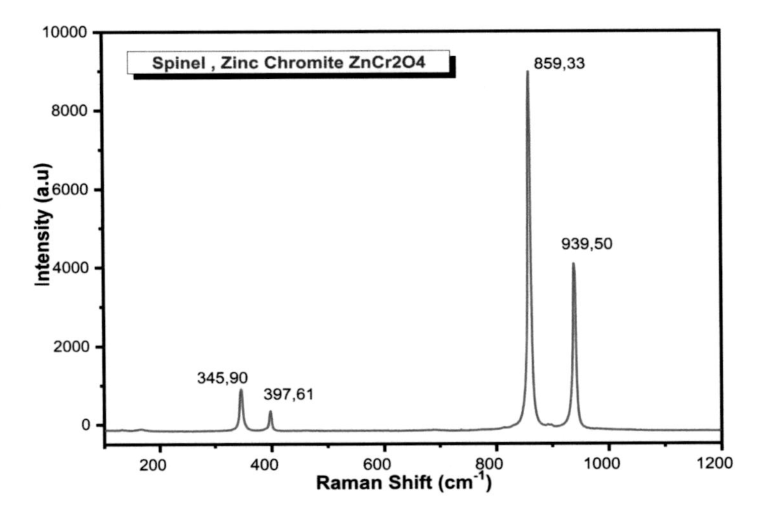

Fig. 11.6 Spinel, Zinc Chromite $ZnCr_2O_4$

11.1.5 KDP

KDP is considered the prototype of isomorphic compounds with the generic formula MY_2XO_4 (M = K, NH, Rb, and Cs), (Y = H, D), (X = P, As). This material exhibits various phases depending on the temperature. For example, when cooled below room temperature, it shows spontaneous polarization, with a critical temperature close to 122 K, and monoclinic crystalline ordering [23]. Above this temperature, KDP undergoes an order-disorder phase transition, where the two configurations of the disordered PO4 tetrahedra correspond to the two orientations of the permanent dipoles. The crystalline ordering adopted in this phase is tetragonal with space group $I4^{-2}$ m. In the high-temperature (HT) region, it presents a series of phases associated with structural and molecular order, the origin of which is a matter of discussion. It is reported in the literature that near 180 °C, it undergoes a phase transition from tetragonal II to monoclinic II'. For this solid-solid phase transition [23, 24].

KDP is widely used as an electro-optic modulator, Q-switch, and high-power laser frequency conversion material due to its excellent performance as an active element in devices exhibiting piezoelectric, ferroelectric, electro-optic, and nonlinear optical responses [25] (Fig. 11.7).

The most representative peaks of the spectrum are marked by: ν_2 modes associated with vibrations in the PO4 bond's plane (951–973 cm^{-1}); ν_4 modes linked to the collective movements of protons in the c-plane, which are sensitive to the presence of other bonds besides hydrogen bonds; ν_1 modes associated with PO4 stretching vibrations; and ν_3 modes, which correspond to non-symmetric vibrations that overlap with ν_1 modes near room temperature. As the material's temperature increases, the Raman spectra reveal the emergence of new bands. These bands are associated with changes in the vibrational ordering of crystalline KADPx.

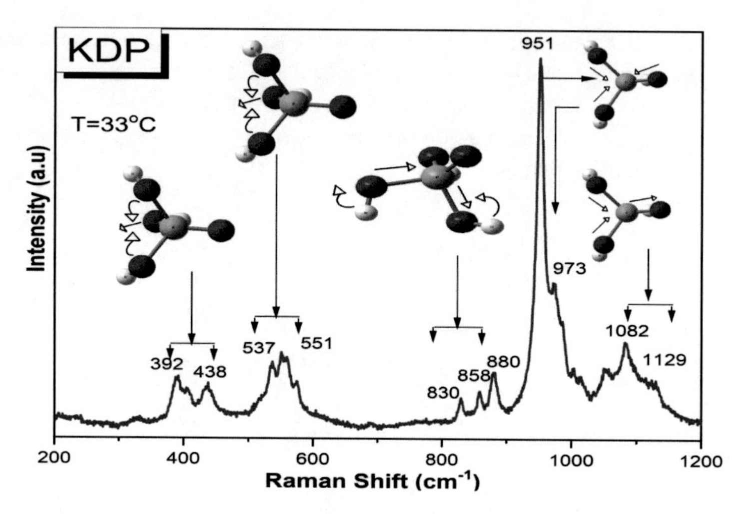

Fig. 11.7 KDP

The shift and presence of these new bands are associated with changes in vibration modes, which may correlate with the material adopting a new structural organization [26]. This new ordering suggests that the internal vibrations of the PO4 ions have been modified, presumably due to changes in their environment. Specifically, the new bands might highlight modes associated with ν_s(P-O-P) and ν_s(O-P-O), located at 537–551 cm^{-1} and 1082 cm^{-1}, respectively.

This new molecular ordering is added to the initial tetragonal phase. The coexistence of these orderings encompasses the phase associated with low temperatures, and the transition to the new organization occurs through a dynamic and progressive process, which substantially depends on temperature and the duration of isothermal conditions. These observations support the proposal that a polymerization process occurs within the material. This leads us to infer that the polymerization times or the formation of $\mathrm{P_n}O_{3n+1}$ chains are relatively long. It is important to note that even after several minutes, the phase associated with low temperatures is still present in a small percentage and can likely be recovered due to KADPx's tendency to absorb environmental moisture, which may prevent or disrupt polymerization.

11.1.6 Gallium Arsenide

Gallium arsenide (GaAs) is a semiconductor with significant applications in electronics and optoelectronics, particularly in sensors and light-emitting devices. It has been studied extensively over the years. Raman spectroscopy is a valuable tool for investigating the vibrational properties of GaAs, as it provides information about the crystal structure, quality, and the presence of defects or impurities. The positions, intensities, and widths of the Raman peaks in GaAs depend on temperature. As the temperature increases, the phonon peaks generally broaden and shift to lower frequencies due to anharmonic effects. Therefore, Raman spectroscopy is used for in-situ monitoring of the manufacturing process and optimization of material properties during device fabrication and quality control.

Crystal Structure and Phonon Modes: GaAs have a zinc blende crystal structure, which is a face-centered cubic lattice. The vibrational modes observed in Raman spectroscopy of GaAs are the longitudinal optical phonon (LO) and the transverse optical phonon (TO), both of which are Raman-active modes.

- The longitudinal optical phonon (LO) mode in GaAs is typically observed around 292–296 cm^{-1}.
- The transverse optical phonon (TO) mode is usually observed around 267–270 cm^{-1}.

The positions of these peaks can vary slightly depending on factors such as temperature, strain, and doping. Dopants in GaAs can shift the LO and TO phonon peaks. For example, n-type doping can shift the LO phonon peak to higher frequencies due to plasmon-phonon coupling, as n-type doping aims to increase the majority of electron carriers. When stress occurs in the crystal, such as in heterojunctions where GaAs is the substrate or part of the substrate, the stress produced by lattice mismatches can cause shifts in the Raman peaks. Tensile stress typically shifts the peaks to lower frequencies, while compressive stress shifts them to higher frequencies, similar to the behavior observed in the band gap of the semiconductors.

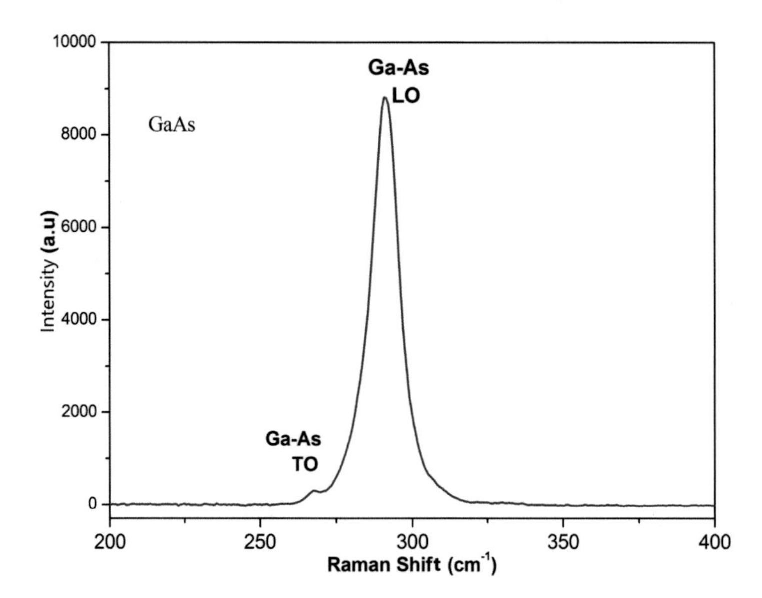

Fig. 11.8 Gallium Arsenide

One technique that provides additional information is Resonant Raman Scattering. Resonant Raman scattering can occur in GaAs when the incident laser energy is close to the electronic band gap (1.42 eV). This can enhance the intensity of specific Raman peaks and provide further insights into the electronic structure. In addition to the intense LO and weaker TO phonon peaks, additional peaks can be observed depending on the specific fabrication conditions of the sample, such as doping levels or the presence of surface oxides (Fig. 11.8).

11.1.7 GaAs/GaAs. GaAs Film Growth on a GaAs Substrate (i.e., GaAs/GaAs)

GaAs films were deposited on GaAs substrates using the metal-organic chemical vapor deposition (MOCVD) technique [26]. Raman spectra were obtained at different temperatures using a sample holder with a microheater integrated inside the system, and temperature changes were performed at a rate of 10 °C/min (±1 °C), with a thermal stabilization time of 10 minutes. As a function of temperature, the Raman spectra show a shift toward lower energy values as the temperature increases. The new normal vibration mode that appears and the shift of the central peak is associated with the tensile deformation in the lattice caused by the difference in thermal expansion coefficients between the film and the substrate [27] (Fig. 11.9).

The spectrum shows the peak located at 292 cm^{-1}, assigned to the LO-GaAs mode. This peak shifts to lower energy values as the sample temperature increases.

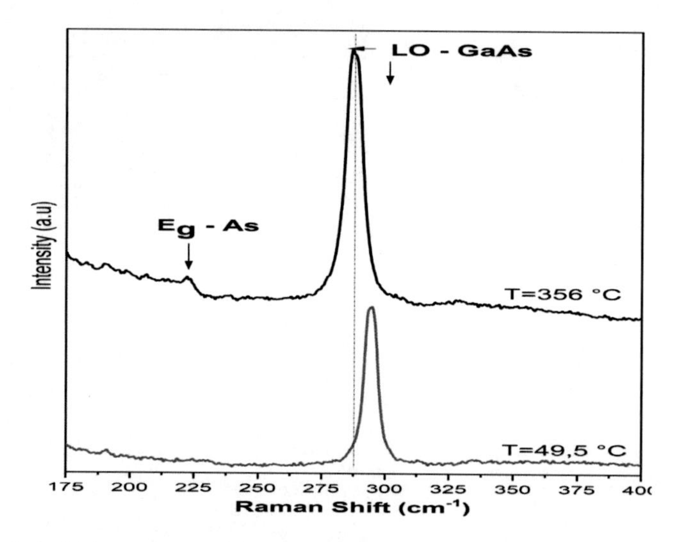

Fig. 11.9 GaAs film growth on a GaAs substrate (i.e., GaAs/GaAs)

A similar behavior is evident for a band around 220 cm^{-1}, associated with arsenic precipitation [28].

From the shift of the frequency peak, it is possible to establish a relationship between the stress percentage and the thermal energy in the GaAs/GaAs system, which can be expressed by the relationship $e = 0.2 \times \Delta\omega$, proposed in the reference [29] (Fig. 11.9).

11.1.8 Calcite

Studies of the calcite ($CaCO_3$) Raman spectrum show several peaks (Fig. 11.10). The primary Raman peaks associated with the carbonate (CO_3^{2-}) group in calcite are:

- 1085 cm^{-1} is the most intense and prominent peak, corresponding to the symmetric stretching vibration (ν_1) of the carbonate ion (CO_3^{2-}).
- 711 cm^{-1} peak corresponds to the in-plane bending mode (ν_4) of the carbonate ion.
- 282 cm^{-1} peak is due to the lattice vibration mode of the calcium ions relative to the carbonate group.
- 1436 cm^{-1}, peak corresponds to the asymmetric stretching vibration (ν_3) of the carbonate ion.

These peaks are characteristic of calcite and can be used to differentiate it from other polymorphs of calcium carbonate, such as aragonite and vaterite.

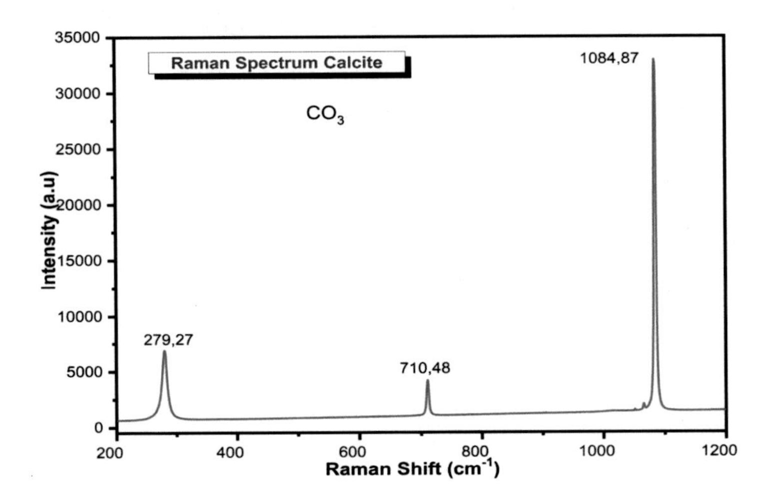

Fig. 11.10 Calcite

11.2 Organic Materials

11.2.1 Cellulose

Cellulose is a complex carbohydrate or polysaccharide that is a key structural component of plant cell walls. It is composed of long chains of glucose molecules linked together by $\beta(1-4)$ glycosidic bonds, forming a linear, rigid structure. This structure allows cellulose to form strong fibers that contribute to plant cells' mechanical strength and rigidity. Cellulose is Earth's most abundant organic polymer, making up about 33% of all plant matter, including 90% of cotton and 50% of wood. In plants, cellulose fibers are bundled together to form microfibrils, which provide tensile strength to the cell walls, allowing plants to stand upright and resist external forces.

Cellulose has numerous industrial applications. It is used to produce paper and cardboard, textiles (such as cotton and linen), and various cellulose derivatives like cellulose acetate (used in making films and plastics) and carboxymethyl cellulose (used as a thickener in food and cosmetics). Due to its highly ordered structure and extensive hydrogen bonding, cellulose is insoluble in water and most organic solvents, making it a challenging material to process (Fig. 11.11).

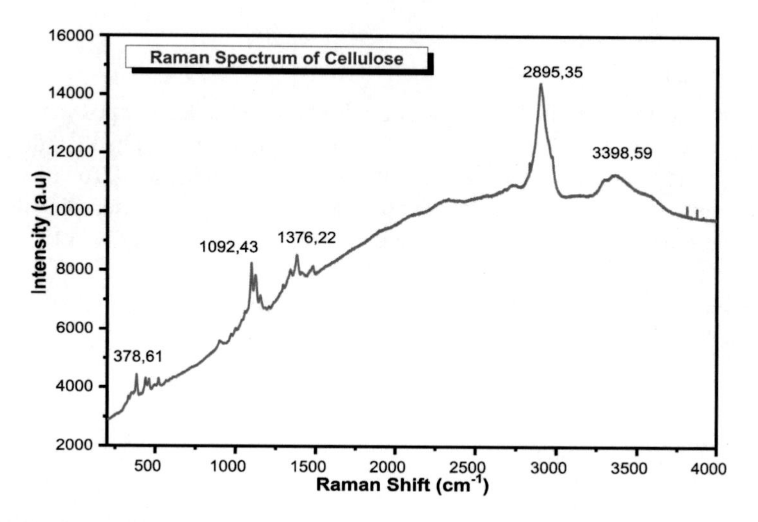

Fig. 11.11 Cellulose

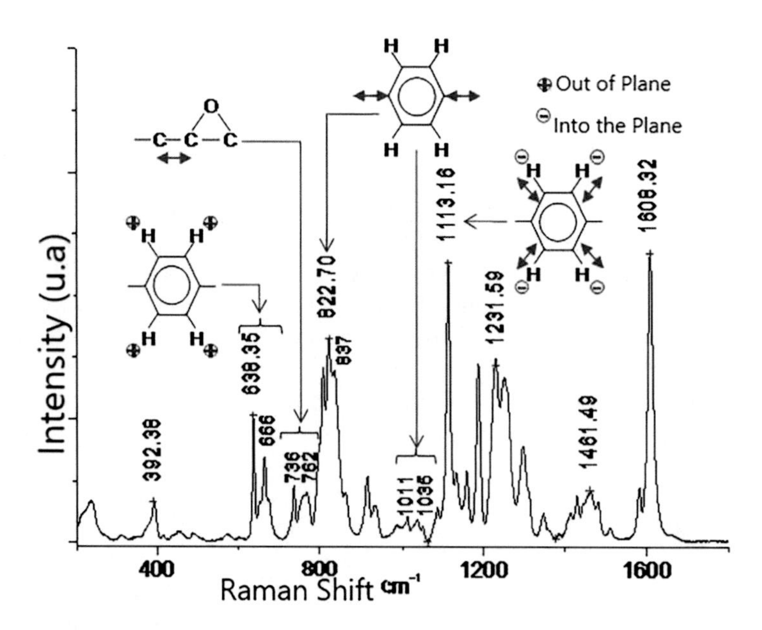

Fig. 11.12 Epoxy Resin

11.2.2 Epoxy Resin

Several peaks are associated with the normal vibration modes of epoxy rings (see Fig. 11.12. Some of these peaks have greater intensity than others, and their decrease with conversion indicates the progress of the cross-linking reactions. The peaks that can be considered potential candidates for monitoring curing, based on different vibrational modes of the epoxy group, are 736, 915, 933, 1131, 1249–1252, 1407, 2713, 2760, and 3006 cm^{-1}.

The most complete list of Raman peaks for epoxy resin is shown in the following table (Tables 11.2 and 11.3):

11.2.3 TMAB Diamine

Figure 11.13 shows the Raman peaks associated with diamine. This table explains the relationship between the normal modes and the motion of the respective atoms (Fig. 11.13 and Table 11.4).

Table 11.2 Epoxy Resin

Raman Shift (cm^{-1})	Vibrational Mode	Description
800–900	C–H out-of-plane deformation	Associated with aromatic rings in the epoxy resin
1250–1300	C–O–C stretching	Vibrations within the epoxy ring
1600	C=C stretching	Aromatic C=C stretching vibrations
2900	C–H stretching	General C–H stretching vibrations in organic compounds
3050	Aromatic C–H stretching	Specific to aromatic rings in the epoxy resin

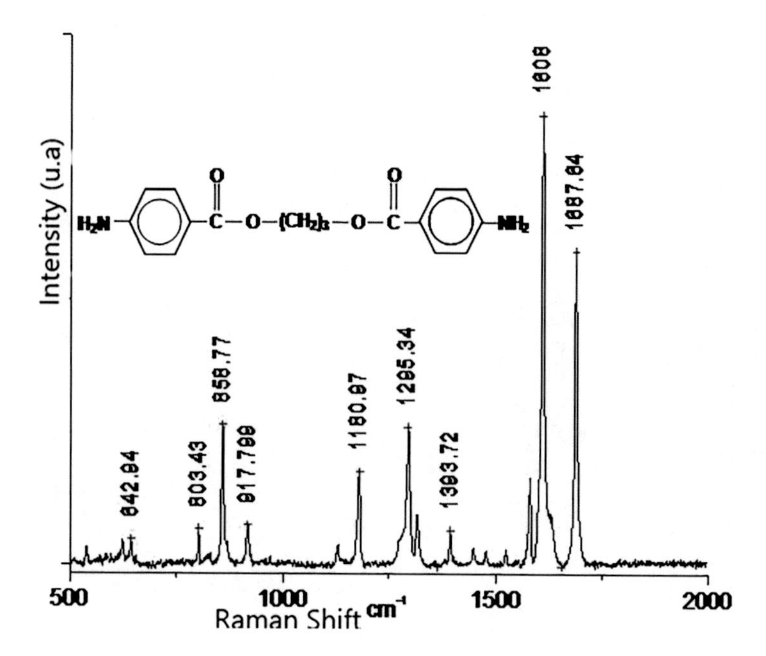

Fig. 11.13 TMAB Diamine

11.3 Other Materials

11.3.1 Hair

The Raman spectrum of hair is complex due to its composition, which includes proteins (mainly keratin), lipids, and pigments (such as melanin). Here are some key Raman shifts typically observed in the Raman spectrum of hair (Fig. 11.14):

Table 11.3 Raman peaks for epoxy resin, w: Weak; (b) m: Medium; (c) s: Strong; (d) vs: Very Strong. References [30–33, 33–37]

Vibrational modes	Peaks cm^{-1}
ν(C–H): Deformation of the carbon-hydrogen bonds outside the plane of the aromatic ring.	641(s), 667(m), 640(m)
ν(C–C): Vibration of the carbon-carbon bonds adjacent to the epoxy ring.	736(s), 762(m)
ν(CH_2): In-plane deformation of the CH_2 group in the epoxy ring.	846(m), 840-853
ν(C–O–C): Vibration of the epoxy ring.	908(m), 924(s), 912-920, 914(m), 1211(s)
ν(C−): Vibration of the substituted bond in the aromatic ring.	1010(w), 1036(w)
ν(C–H): In-plane deformation of the carbon-hydrogen bonds in the aromatic ring.	1010(w)
ν(C–H): Vibration of the carbon-hydrogen bonds in the aromatic ring and in-plane deformations.	1113(s)
ν(CH_2): Deformation of the carbon-hydrogen bonds in the epoxy ring (wagging).	1130(m)
ν(C–CH_3): Vibration of the carbon-carbon bonds (of the methyl group) and in-plane deformation of the methyl groups.	1186(s)
ν(C–H): In-plane deformation of the carbon-hydrogen bonds in the aromatic ring.	1190(s)
ν(C–O): Vibration of the carbon-oxygen bond with the phenyl group and with the carbon of the CH_2 group.	1232(s)
ν(C–O): Vibration of the carbon-oxygen bond with the carbon of the CH_2 group.	1248(s), 1256(s)
ν(C–O–C): In-plane deformation of the epoxy ring.	1259(s), 1265(m), 1270(m)
ν(CH_2): Deformation of carbon-hydrogen bonds in the epoxy ring (torsion)	1407(m)
ν(C–H): Deformation of carbon-hydrogen bonds in and out of the plane in CH_2 groups.	1450(m), 1464(m), 1477(w)
ν(CC): Vibration of carbon-carbon bonds in and out of the plane of the aromatic ring; deformation of CH bonds.	1584(m), 1583(m), 1604(m), 1602(s), 1612(m), 1614(s)
ν(–CH–): Symmetrical vibration of the bonds in the CH group of the oxirane ring.	2716(w)
ν(–CH–): Asymmetrical vibration of the bonds in the CH group of the oxirane ring.	2763(w)
ν(–O–CH_2–): Vibration of the CH_2 group in the ether.	2833–2838, 2840(m)
ν(–O–CH_2–): Symmetrical vibration of the CH_2 group in the ether.	2873(m), 2875(m), 2873-2875
ν(–O–CH_2–): Asymmetrical vibration of the CH_2 group in the ether.	2931(m), 2922-2925, 2919(m)

Table 11.3 (continued)

Vibrational modes	Peaks cm^{-1}
ν(C–H): Asymmetrical vibration of the carbon-hydrogen bond in methyl groups.	2971(m), 2944–2953
ν(–CH_2–): Symmetrical vibration of the CH_2 groups in the oxirane rings.	3004(m), 3010(m), 3002–3005,3000(m)
ν(C–H): Vibration of the carbon-hydrogen bonds in the aromatic ring with deformation in and out of the plane.	3055(s), 3062(s), 3069(s), 3070(s), 3038–3046, 3068–3073, 3071(s)

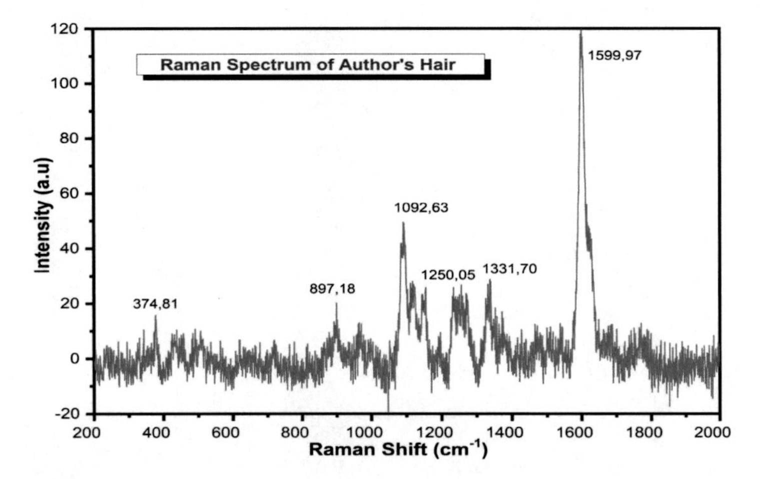

Fig. 11.14 Raman Spectrum of Author's Hair

- 500–600 cm^{-1}: Peaks associated with keratin disulfide bonds (S–S).
- 1000–1030 cm^{-1}: Peaks associated with C–C stretching vibrations.
- 1200–1300 cm^{-1}: Peaks associated with the Amide III band, related to C–N stretching and N–H bending.
- 1440–1470 cm^{-1}: Peaks associated with CH_2 and CH_3 bending vibrations.
- 1500–1600 cm^{-1}: Peaks associated with aromatic ring stretching, often attributed to melanin.
- 1600–1650 cm^{-1}: Peaks associated with the Amide I band, related to C=O stretching in proteins.
- 2850–2950 cm^{-1}: Peaks associated with CH_2 and CH_3 stretching vibrations in lipids and proteins.

These peaks are related to the molecular vibrations within hair's proteins, lipids, and pigments. The exact Raman shifts and intensities can vary based on hair color, treatment, and overall health. However, it is essential to consider that the laser power

Table 11.4 TMAB Diamine, w: Weak; (b) m: Medium; (c) s: Strong; (d) vs: Very Strong

Normal modes	Raman peaks cm^{-1}
ν(H): Vibration of hydrogen bonds in the plane of the aromatic ring	622(m), 622(m)
ν(¿?): Unidentified vibrational mode	636(m)
ν(CC): Vibration of carbon-carbon bonds in the central chain of the material	600–1300(m)
ν(C–O–C) (Symmetrical): Vibration of carbon-oxygen-carbon bonds (Symmetrical)	800–970(m)
ν(C–O–C) (Asymmetrical): Vibration of carbon-oxygen-carbon bonds (Asymmetrical)	1060–1150(w)
ν(CH): In-plane deformations of carbon-hydrogen bonds	1154(m), 1155(m), 1156(m), 1181(m), 1182(m), 185(m), 1198(m), 1201(m), 1205(m)
ν(CH_2) (Twisting): Oscillation in and out of the plane (Twisting)	1294(m), 1300(s), 1303(m)
ν(CH_2) (Wagging): Oscillation in and out of the plane (Wagging)	1425(vw), 1448(m, s), 1454(m)
ν(N–H): Vibration of nitrogen-hydrogen bonds in amine groups (Symmetrical and Asymmetrical). Primary amines	1624(m)
ν(C=O): Carbon-oxygen double bond oscillation	1680–1820(m)
ν(CH): In-plane deformations of carbon-hydrogen bonds	2849(m), 2906(m)
ν(O–CH_2): Vibrations of oxygen-CH_2 bonds	2833(m)
ν(CC): Vibration of carbon-carbon bonds in and out of the aromatic ring plane, deformation of CH bonds	3055(s), 3062(s)
ν(–H): Vibration of hydrogen bonds	3000–4000
ν(N–H): Vibration of nitrogen-hydrogen bonds in amine groups (symmetrical and asymmetrical). Primary and secondary amines	3330–3400, 3300–3600, 3304–3366, 3340–3361, 3404–3422, 3367, 3454

can damage the keratin and remove this protective layer. Typically, the power is kept lower than five mW per micrometer.

11.3.2 Bone

The Raman spectrum of bone is complex and primarily reflects its mineral and organic components. Bone is composed mainly of hydroxyapatite (a form of calcium phosphate) and collagen (a type of protein). Here are some fundamental Raman shifts typically observed in the Raman spectrum of bone (Fig. 11.15):

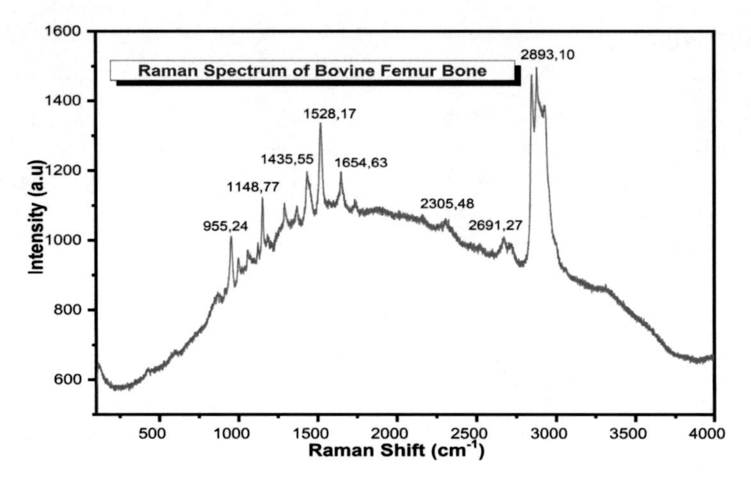

Fig. 11.15 Bovine Femur Bone

- 430–460 cm^{-1}: Peaks associated with phosphate (PO_4^{3-}) bending modes.
- 580–610 cm^{-1}: Peaks associated with carbonate (CO_3^{2-}) bending modes.
- 960 cm^{-1}: The most prominent peak, associated with the symmetric stretching of phosphate (ν_1 PO_4^{3-}), a key component of hydroxyapatite.
- 1070 cm^{-1}: Peaks associated with carbonate CO_3^{2-}) substitution in the hydroxyapatite lattice.
- 1240–1280 cm^{-1}: Amide III band, associated with collagen, reflecting C–N stretching and N–H bending.
- 1440–1470 cm^{-1}: CH_2 bending vibrations associated with collagen.
- 1660–1690 cm^{-1}: Amide I band, associated with collagen, reflecting C=O stretching in proteins.
- 2850–2950 cm^{-1}: CH_2 and CH_3 stretching vibrations, primarily related to lipids and proteins.

These Raman peaks provide insights into the bone's mineral content (primarily hydroxyapatite) and its organic matrix (mainly collagen). The exact Raman shifts can vary slightly depending on the bone's composition, health, and treatment. For the synthesis of hydroxyapatite using phosphates and other reagents such as carbonates, it is important to verify the presence of contamination or other residual materials (Fig. 11.15).

11.3.3 Snail Shell

A Raman spectrum of a snail obtained from Corpus Christi, Texas, shows that the snail's shell has distinct regions, primarily two: one white and one black. These regions have been measured. Any biological sample provides a detailed vibrational fingerprint that reflects its molecular composition. The spectrum typically shows peaks corresponding to various biomolecules, such as proteins, lipids, carbohydrates, and nucleic acids.

In the case of a snail, the Raman spectrum might highlight:

- Proteins: Peaks around 1000–1700 cm^{-1}, related to amino acid vibrations and protein structures.
- Lipids: Characteristic peaks near 2850–3000 cm^{-1}, corresponding to CH stretching vibrations in fatty acids.

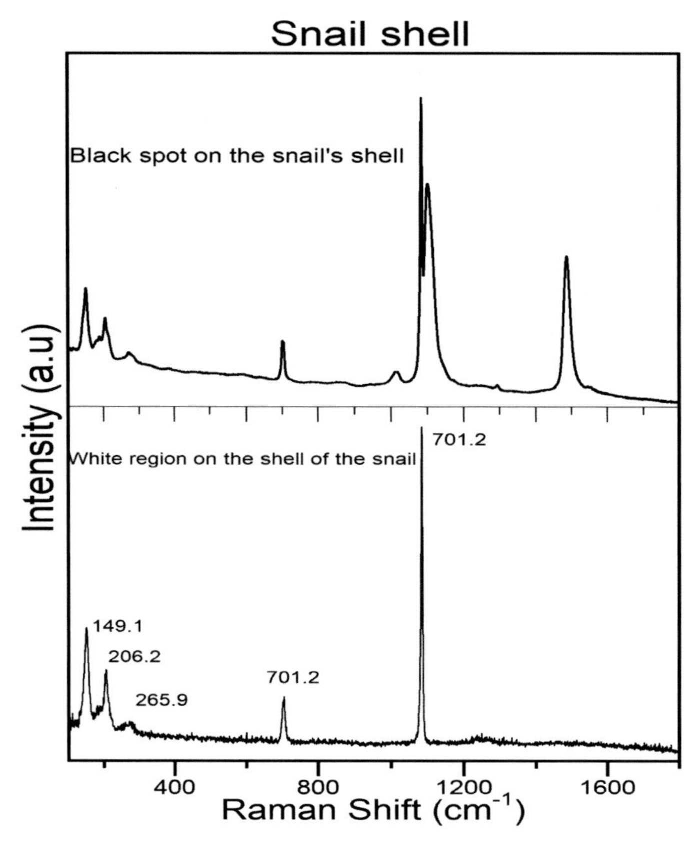

Fig. 11.16 Snail Shell

- Carbohydrates: Peaks in the range of 800–1200 cm^{-1}, associated with the vibrational modes of sugar rings and other carbohydrate structures.
- Calcium Carbonate: Peaks around 1085 cm^{-1}, due to the presence of calcium carbonate in the snail's shell.

The spectrum can vary depending on the part of the snail being analyzed (e.g., shell or soft tissues) and specific conditions (such as hydration or the presence of other substances) (Fig. 11.16).

11.3.4 Water

The normal vibrational modes of the water molecule can be determined by Raman spectroscopy. When we say that a Raman mode is active, it means it can be detected using this technique. As an independent molecule in the gas phase, water has three fundamental vibrational modes, which can be classified as follows:

- **Symmetric Stretching** (ν_1): In this mode, the two O-H bonds stretch symmetrically, meaning both hydrogen atoms move away from and toward the oxygen atom simultaneously. This mode is Raman active and typically appears around 3652 cm^{-1} in the gas phase.
- **Asymmetric Stretching** (ν_3): In this mode, one of the two O-H bonds stretches while the other compresses, creating an asymmetry in the molecular vibrations. This mode is also Raman active and usually appears around 3756 cm^{-1} in the gas phase.
- **Bending Mode** (ν_2): In this mode, the H–O–H bond angle flexes, with the two hydrogen atoms moving closer together and farther apart, changing the angle between the O–H bonds. The bending mode is Raman active and typically appears around 1595 cm^{-1} in the gas phase.

These vibrational modes are crucial for understanding the molecular structure and behavior of water. Specific Raman shifts can vary slightly depending on the phase (gas, liquid, or solid) and environmental conditions, such as temperature, pressure, impurities, or if the water is diluted with another liquid.

When water molecules cluster to form liquid water, the Raman spectrum becomes more complex compared to its gas phase due to hydrogen bonding and interactions between adjacent water molecules. In liquid water, the vibrational modes broaden due to disorder, and the peaks are shifted compared to the gas phase. All three vibrational modes of liquid water remain Raman active and are located at:

- O–H Stretching (ν_1 and ν_3): The symmetric (ν_1) and asymmetric (ν_3)) stretching modes in liquid water overlap and form a broad band in the Raman spectrum, typically located between 3000 and 3700 cm^{-1}. This band is broadened by the extensive hydrogen bonds in liquid water, which cause a distribution of O-H bond lengths and strengths, leading to the broadening and shifting of the peaks.

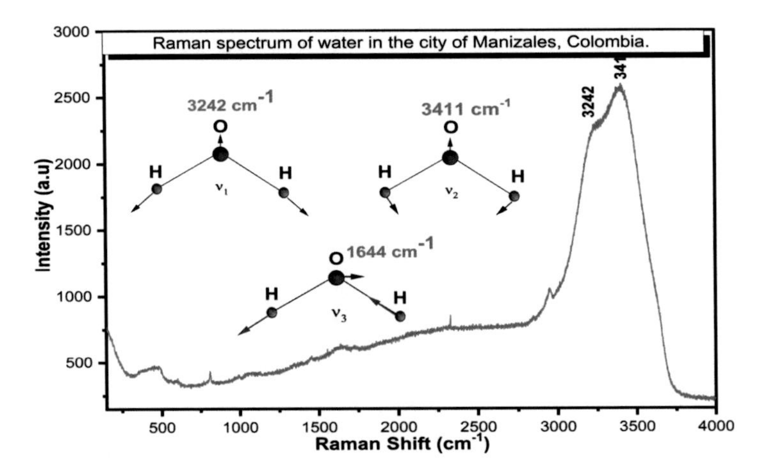

Fig. 11.17 Water

- **H–O–H Bending Mode** (ν_2): This bending mode appears as a more distinct or defined peak in the Raman spectrum, usually located around 1600 to 1650 cm^{-1}. This mode is less affected by hydrogen bonding than the stretching modes, so it remains relatively sharp, though still broader than in the gas phase.
- **Low-Frequency Modes:** Below 400 cm^{-1}, additional features can be observed, associated with the collective motions of water molecules, such as translational and librational (rotational) motions. These modes are significantly influenced by hydrogen bonding and the dynamic structure of the liquid.-

In summary, the Raman spectrum of liquid water is characterized by a broad O–H stretching band (3000–3700 cm^{-1}) and a more defined H–O–H bending mode (around 1600–1650 cm^{-1}), with additional low-frequency modes below 400 cm^{-1}. The broadening of the peaks in the O–H stretching region reflects the hydrogen bonding interactions that are a hallmark of the liquid state. Collective modes arise when individual molecules form clusters in the liquid or solid phase, and these collective modes are found in the low-frequency bands (Fig. 11.17).

11.3.5 PEO

Polyethylene oxide (PEO), also known as polyethylene glycol (PEG) in its lower molecular weight form, exhibits characteristic Raman peaks that can be used to identify and analyze the material. PEO has several distinct peaks in its Raman spectrum, and some of the important Raman peaks for polyethylene oxide are:

C–C stretching vibrations: 1065 cm^{-1} is one of the most prominent peaks associated with the C–C stretching vibrations.

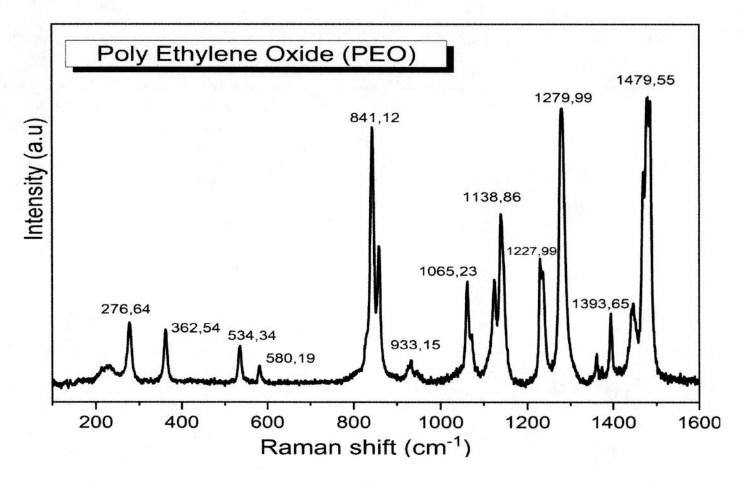

Fig. 11.18 Raman Shift normal vibrational modes for PEO

C–O–C stretching vibrations: 841.12 cm^{-1}, this peak corresponds to the symmetric stretching of the C–O–C group. 933.15 cm^{-1}, this peak is related to the asymmetric stretching of the C–O–C group.

C–H bending vibrations: 1279.99 cm^{-1}, this peak represents the bending vibrations of the C–H bond.

CH_2 rocking: 1479.55 cm^{-1}, this peak corresponds to the CH_2 rocking mode.

CH_2 twisting: 1365 and 1393.65 cm^{-1}, these peaks are associated with the twisting of the CH_2 group. These are some of the typical Raman peaks observed in the spectrum of polyethylene oxide. These peaks' exact position and intensity can vary depending on the PEO sample's molecular weight, crystallinity, and physical form (Fig. 11.18).

The Table 11.5 summarizing the key Raman peaks for polyethylene oxide (PEO).

The following table lists the most important Raman peaks commonly observed in the spectrum of polyethylene oxide, along with their associated vibrational modes and descriptions. The exact positions and intensities of these peaks may vary depending on factors such as the molecular weight and crystallinity of the PEO sample.

11.3.6 Styrofoam

The Raman shift of normal vibrational modes for Styrofoam (polystyrene) typically exhibits several characteristic peaks. Here are some of the key Raman shifts associated with Styrofoam:

- 620–650 cm^{-1}: Peaks associated with aromatic ring deformation.

Table 11.5 Raman peaks for polyethylene oxide (PEO)

Raman shift cm^{-1}	Vibrational mode	Description
841.12	C–O–C symmetric stretching	Symmetric stretching of the ether group
933.15	C–O–C asymmetric stretching	Asymmetric stretching of the ether group
1065.23	C–C stretching	Stretching of the carbon-carbon bonds
1279.99	C–H bending	Bending vibrations of the methylene group
1393.65	CH_2 twisting	Twisting motion of the methylene group
1479.55	CH_2 rocking	Rocking motion of the methylene group

- 1000–1030 cm^{-1}: Peaks associated with in-plane ring stretching (breathing mode).
- 1150–1200 cm^{-1}: Peaks associated with C–H in-plane bending.
- 1200–1300 cm^{-1}: Peaks associated with C–H bending modes.
- 1600–1650 cm^{-1}: Peaks associated with aromatic C=C stretching.

These peaks correspond to the vibrational modes of polystyrene's aromatic rings and carbon-hydrogen bonds. The exact position of these peaks can vary slightly depending on the specific sample and measurement conditions (Fig. 11.19).

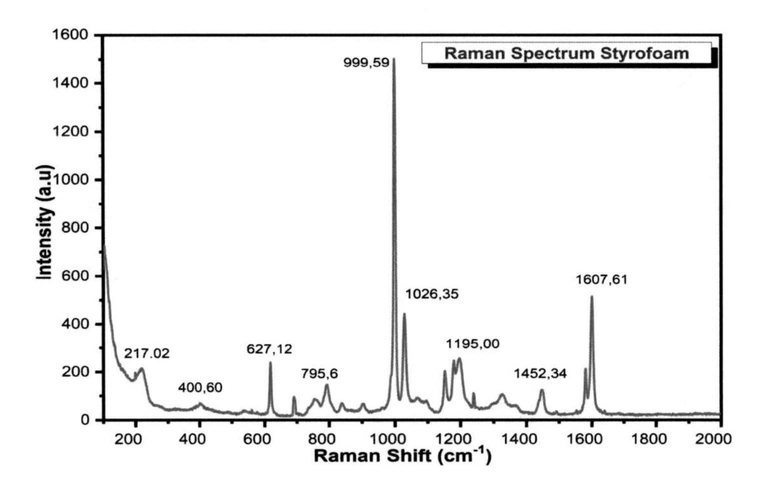

Fig. 11.19 Raman Shift of normal vibrational modes for Styrofoam

11.3.7 DNA

In 1950 and 1951, Maurice Wilkins, Ray Gosling, and Rosalind Franklin began studying the structure of deoxyribonucleic acid (DNA). X-ray diffraction techniques demonstrated the formation of fibers with helical structures, though obtaining detailed information about the helices is required. These findings allowed them to classify the structures into two forms, A and B, depending on the humidity of the surrounding medium. Meanwhile, the work of Francis Crick and James Watson, using Franklin's data and their analyses based on recent findings, led to the development of the first accurate model of DNA's structure. The results were published in Nature in 1953 in articles by Watson and Crick and by Wilkins, Franklin, and Gosling. The researchers were awarded the Nobel Prize in Physiology or Medicine in 1962 for their work [38].

The hereditary characteristics are transmitted by the DNA macromolecule, which is considered, from a structural point of view, as a skeleton formed by two threads joined by alternating bases in a specific sequence. These threads are structured as a polymeric chain formed by monomeric units called nucleotides, which comprise a phosphate group, a pentose sugar (deoxyribose), and a nitrogenous base—the phosphate group and the deoxyribose act as a backbone or support for the nitrogenous bases. The nitrogenous bases that join these threads are aromatic radicals that can be a purine (adenine, guanine) or a pyrimidine (cytosine, thymine). The structure of DNA has symmetrical characteristics, among which are the rings formed by the carbons. These rings can be single or double: thymine and cytosine have a single-ring structure known as pyrimidine bases, while adenine and guanine, which have a double-ring structure, are known as purine bases. This is important in the interaction due to the vibrational resonance of DNA. Part of this symmetry in the ring type allows the complement between their structures. The genetic transmission system indicates that adenine can only bind to thymine, while cytosine can only bind to guanine. This union is carried out by hydrogen bonds, where thymine and adenine share two hydrogen bonds and cytosine and guanine share three.

One of the DNA structures that plays a vital role in the link or interaction with some nanoparticles is the phosphate group since it forms a link between nucleotides using a phosphodiester bond. Each aromatic or deoxyribose (sugar) is linked to the phosphate group and the nitrogenous base, so the difference between nucleotides lies in the base to which they are connected. Therefore, in a DNA chain, the essential characteristic is the sequence of the nitrogenous bases, so its fingerprint is the order of the nitrogenous bases within the chain. How the four bases are distributed or organized internally along the chain is what allows the decoding of genetic information. The phosphate linked to the sugar is organized in a way known as plectonemic or spiral staircase coiling.

In the IUPAC system (see the book by Nakamoto et al. [38]), the organization or arrangement of carbon and nitrogen atoms in the rings of the DNA structure is numbered according to this system. Thus, each functional group (methyl, amino, or oxo) is associated with the number or position of the atom in each of the rings to which it is linked. The structure formed by one of the bases and deoxyribose is called a nucleotide. As for the phosphor oxide group (PO_2^-), this is formed by semi-double bonds (P–O), where the negative charge of the oxygen atoms is shared.

The normal vibration modes of the DNA molecular structure can be determined by infrared or Raman spectroscopy. The peaks assigned to the expected vibration modes of the DNA-B structure in solution are shown in Table 11.6. Some characteristic peaks include the band at 1670 cm^{-1}, corresponding to the exocyclic carboxyl (C=O) in its base bond deformation mode; a peak at 1578 cm^{-1}, attributed to the compression overlap of the bonds in guanine and adenine; and a band at 1489 cm^{-1}, assigned to the intense polarization of purines. There are two scissoring modes assigned to CH_2 bonds at 1420 cm^{-1} and four intense Raman bands at 1254, 1302, 1340, and 1375 cm^{-1}, which can be attributed to the vibration plane of the residual base rings. The peak at 1092 cm^{-1} is assigned to PO_2^-, a strongly polarized molecule associated with symmetric deformations. The bands located at 895 and 920 cm^{-1} are linked to the deformation vibration of the deoxyribose ring bonds. In comparison, the modes in the range of 750–850 cm^{-1} correspond to the depolarization modes of phosphodiesters. Other bands include 683 (G), 728 (A), 749 (T), and 782 (C) cm^{-1}, which are assigned to the "breathing" motion of the residual base rings, where G and C exhibit high and low polarization values, respectively. The normal vibration modes for the conformational structures of DNA-A and DNA-B are shown in Table 11.7.

For the preparation of the DNA sample in the Raman measurements, the supernatant was removed to allow the precipitate to be resuspended in ethylenediaminetetraacetic acid (EDTA, $C_{10}H_{16}N_2O_8$). This step enables the chelation of Ca^{2+} ions, inhibiting DNAse activity and preventing DNA degradation. The DNA fibers were precipitated by adding 99.7% ethanol at −20 °C, then dried, and finally resuspended in TE buffer (10 mM Tris + 1 mM EDTA) at room temperature. The TE buffer helps control the pH and temperature, ensuring proper dispersion of the DNA and preventing its destruction during laser exposure.

The figure shows spectral lines that correspond to those reported in the literature, associated with the vibrational modes of the internal and peripheral components of the DNA molecule. The range between 200 and 4000 cm^{-1} spectrum shows thirteen representative lines, each indicating a normal vibration mode associated with the molecular structure components and their interactions (Fig. 11.20).

- The stretching oscillations of the nitrogenous bases, cytosine, and thymine rings are manifested in the band at 786 cm^{-1} [40].
- The intensity band at 863 cm^{-1} is not clearly assigned. Some studies suggest it is due to carbohydrate vibrations through C–C or C–O–C bond stretching [40], while others attribute it to stretching oscillations of the deoxyribose ring. These oscillations indicate how hydrogen bonds between the polar groups of the DNA sugar-phosphate backbone and water molecules are modified. This peak strongly

Table 11.6 Raman peaks for B-DNA in solution [38]

Position of peak	Assignment
1670	C=O stretch (T) (T)
1578	Ring mode (A, G)
1498	Ring mode (G, A)
1420	CH_2 scissoring
1375	Ring mode (T)
1340	Ring mode (G, A)
1302	Ring mode (A, C)
1254	Ring mode (C, T)
1092	PO^{-2} symmetric stretch
923	Deoxyribose
900	Deoxyribose
832	O–P–O stretch
792	O–P–O stretch
782	Ring oscillation (C)
749	Ring oscillation (T)
728	Ring oscillation (A)
683	Ring oscillation (G)
670	Ring mode (T)
497	PO^{-2} scissoring

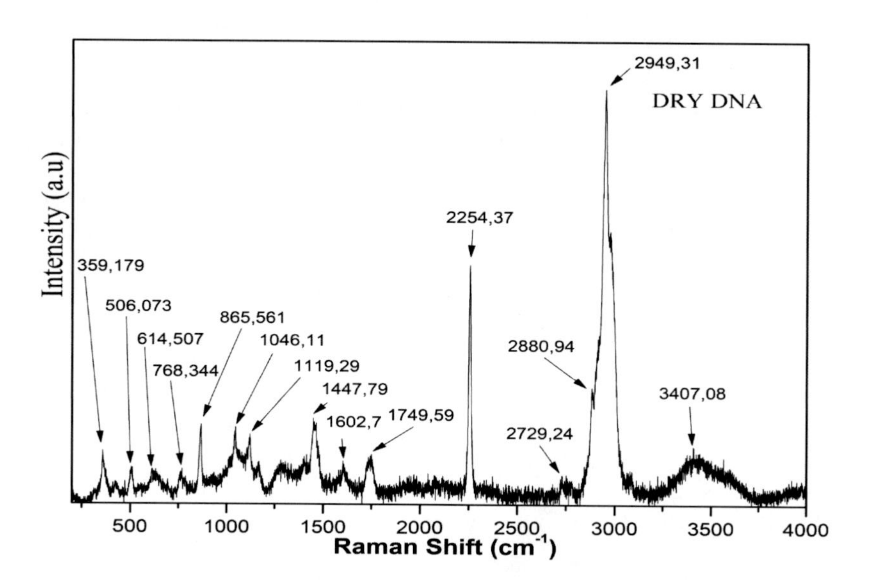

Fig. 11.20 Raman spectrum for the dried DNA sample on a glass substrate [45, 46]

Table 11.7 Normal vibration modes and their respective peaks reported in the literature for the conformational structures of DNA-A and DNA-B [39]

N°	Assignment	DNA-A cm^{-1}	DNA-B cm^{-1}
1	PO_2^- scissoring	499	497
2	Ring Stretch (G)	666	682
3	Ring Stretch (A)	731	729
4	Ring Stretcho (T)	753	750
5	Ring Stretch (C)	784	784
6	C–O–P–O–C Stretch	807	792
7	C–O–P–O–C Stretch	–	836
8	Deoxyribose Ring	882	895
9	Deoxyribose Ring	897	920
10	PO_2^- Symmetric Stretch	1101	1093
11	Ring Vibration (C, T)	1257	1257
12	Ring Vibration (A, G)	1339	1339
13	Ring Vibration (T)	–	1375
14	2′CH_2 scissoring	1419	1420
15	5′CH_2 scissoring	1463	1465
16	Ring vibration (G, A)	1486	1487
17	Ring vibration (A, G)	1577	1578
18	C=O stretching (T)	1667	1669

suggests that changes in the surrounding environment can alter the DNA chain, potentially causing denaturation or a transition from B to Z DNA structures [41].

- The band around 1000 cm^{-1} is situated between two broad bands, from 1020 to 1119 cm^{-1}, associated with the amino acid phenylalanine and the symmetric stretching of the PO_2^- group.
- A faint band around 917 cm^{-1} is assigned to vibrations of the DNA molecule's backbone and deoxyribose. The backbone's rigidity indicates that this normal vibration mode is significantly activated.
- The peaks at 1020, 1046, 1119, and 1164 cm^{-1} are associated with deoxyadenosine and deoxyguanosine residues, manifested by the stretching of the C–O bond in the sugar backbone and the deformations of the C–C bond. These peaks are evident when the structure coils or when DNA denatures. They are also observed during symmetric stretching vibrations of the phosphate group (PO_2^-) in the DNA ladder [41], along with coupled vibrations of the cytosine and guanine rings, respectively [42].
- The peak at 1279 cm^{-1} is assigned to N–H and C–N vibrations.
- The peak at 1450 cm^{-1} is attributed to the normal vibration mode of the adenine, thymine, and cytosine rings and deformations in the C–H_2 bond of deoxyribose [41]. An additional peak at 1450 cm^{-1} is also linked to the presence of lipids.

- The representative peaks at 1603 and 1750 cm^{-1} are due to vibrations of Amide II, coupled with in-phase stretching vibrations of the N–H and C–N bonds, and ester oscillations [43, 44].

References

1. Klingshirn C (2010) ZnO: from basics towards applications. Springer Series in Materials Science, vol 120. Springer
2. Pizzini S, Butta N, Narducci D, Palladino M (1989) J Electrochem Soc 136:1945
3. Vargas-Hernández C, Jiménez-García FN, Jurado JF, Henao Granada VV (2008) XRD, μ-Raman and optical absorption investigations of ZnO deposited by SILAR method. Microelectron J 39:1347–1348
4. Vargas-Hernández C (2011) Síntesis y caracterización de películas de ZnO: Depósito por baño químico. Universidad Nacional de Colombia. ISBN 978-958-761-065-9
5. Yang Z, Liu QH (2008) J Mater Sci 43(19):6527
6. Chen Y et al (2005) J Chem Phys 123:134701
7. Cusco R, Alarcon-Llado E, Ibañez J, Artus L, Jimenez J, Wang B, Callahan M (2007) Temperature dependence of Raman scattering in ZnO. Phys Rev B: Condens Matter Mater Phys 75(16):165202, 1–11 (2007)
8. Manjon F, Mari B, Serrano J, Romero A (2005) Silent Raman modes in zinc oxide and related nitrides. J Appl Phys (Melville, NY, US) 97(5):053516, 1–4
9. Scepanovic M, Grujic-Brojcin M, Vojisavljevic K, Bernikc S, Sreckovic T (2010) Raman study of structural disorder in ZnO nanopowders. J Raman Spectrosc 41(9):914–921
10. Serrano J, Romero A, Manjon F, Lauck R, Cardona M, Rubio A (2004) Pressure dependence of the lattice dynamics of ZnO: an ab initio approach. Phys Rev B: Condens Matter Mater Phys 69(9):094306, 1–14
11. Vargas-Hernandez C (1994) Tesis de Magister en Ciencias Físicas. Universidad Nacional, Bogota. I/1994
12. Vargas-Hernández C, Jiménez-García FN, Jurado JF, Henao Granada VV (2008) Comparison of ZnO thin films deposited by three different SILAR processes. Microelectron J 39:1349–1350
13. Haber J (2009) Fifty years of my romance with vanadium oxide catalysts. Catal Today 142:100–113
14. Ban C, Chernova N, Whittingham SM (2009) Electrospun nano-vanadium pentoxide cathode. Electrochem Commun 11:522–525
15. Guerra EM, Silva RG, Mulato M (2009) Extended gate field effect transistor using V2O5 xerogel sensing membrane by sol-gel method. Solid State Sci 11:456–460
16. Fang GJ et al (2000) Oriented growth of electrochromic thin films on transparent conductive glass by pulsed excimer laser ablation technique. J Phys D: Appl Phys 33:3018–3021
17. Benmoussa M et al (1995) Structural electrical and optical properties of sputtered vanadium pentoxide thin films. Thin Solid Films 265:22–28
18. Freyland W et al (1983) Functional relationships in science and technology. Group III: crystal and solid state physics, vol 17. Semiconductors. Springer, New York, Berlin, Heidelberg
19. Ramana CV et al (1997) Spectroscopy characterization of electron-beam evaporated V2O5 thin films. Thin Solid Films 305:219–226
20. Mancic1 LM, Vulic Z, Moral PC, Milosevic O (2003) Sensors 3:415–423
21. Barth TFW, Posnjak E, Krist F (1932) 82:325 8
22. Vargas-Hernández C, Almanza O, Jurado JF (2009) EPR, μ-Raman and crystallographic properties of spinel type $ZnCr_2O_4$. J Phys: Conf Ser 167:012037
23. Blinc R, Dimic V, Lahaynar G, Stepisnik J, Zumer S, Vene N (1968) J Chem Phys 49:4996–5000
24. Ferroelectrics, special issue on KH_2PO_4-type ferro- and antiferroelectrics 71–72 (1987)

25. Kumaresan P, Babu SM, Anbarasan PM (2008) Thermal, dielectric studies on pure and amino acid (L-glutamic acid, L-histidine, L-valine) doped KDP single crystals. Opt Mater 30(9):1361–1368
26. Jurado JF, Vargas-Hernández C, Vargas RA (2012) Raman and structural studies on the high-temperature regime of the KH_2PO_4–$NH_4H_2PO_4$ system. Rev Mex Fis 58:411–416
27. Orani D, Ricci A, Quagliano LG, Sobiesierski Z (1999) Physica B 262(1):775–778
28. Goerigk G, Bedel E, Claverie A, Herms M, Irmer G (2002) Mater Sci Eng B91–92:466–469
29. Leycuras A, Carles R, Freundlich A, Neu G, Verie C (1988) In: Materials research society symposium proceedings, vol 116, pp 251–254
30. Musto P, Abbate M, Ragosta G, Scarinzi G (2007) A study by Raman, near-infrared and dynamic-mechanical spectroscopies on the curing behaviour, molecular structure and viscoelastic properties of epoxy/anhydride networks. Polymer 48:3703–3716
31. Nyquist R (2001) Interpreting infrared, Raman, and nuclear magnetic resonance spectra. Academic Press
32. Overbeke ED, Legras J, Carter RJT, McGrail PT, Carlier V (2001) Raman spectroscopy and DSC determination of conversion in DDS-Cured Epoxy Resin: application to Epoxy-Copolyethersulfone Blends. Appl Spectrosc 55:540–551
33. Riegel B, Kiefer W, Hofacker S, Schottner G (2000) FT-Raman spectroscopic investigations on the organic crosslinking in hybrid polymers. Part I: model reactions of simple epoxides. Appl Spectrosc 54:1384–1390
34. Aust JF, Booksh KS, Myrick ML (1996) Novel In Situ probe for monitoring polymer curing. Appl Spectrosc 50:382–387
35. Chike KE, Myrick ML, Lyon RE, Angel SM (1993) Raman and near-infrared studies of an Epoxy Resin. Appl Spectrosc 47:1631–1635
36. Don TM, Bell JP (1998) Fourier transform infrared analysis of polycarbonate/epoxy mixtures cured with an aromatic amine. J Appl Polymer Sci 69:2395–2407
37. Farquharson S, Carignan J, Khitrov V, Senador A, Shaw M (2007) Development of a phase diagram to control composite manufacturing using Raman spectroscopy. Pers Commun
38. Nakamoto K, Tsuboi M, Strahan GD (2008) Drug-DNA interactions: structures and spectra. Wiley
39. Thomas G Jr, Benevides J, Overman S, Ueda T, Ushizawa K, Saitoh M, Tsuboi M (1995) Polarized Raman spectra of oriented fibers of A-DNA and B-DNA: anisotropic and isotropic local Raman tensors of base and backbone vibrations. Biophys J 68(3):1073–1088
40. Maquelin K, Kirschner C, Choo-Smith LP, Van den Braak N, Endtz HP, Naumann D, Puppels GJ (2002) Identification of medically relevant microorganisms by vibrational spectroscopy. J Microbiol Methods 51(3):255–271
41. Muntean C, Halmagyi A, Puia M, Pavel I (2009) FT-Raman signatures of genomic DNA from plant tissues. Spectroscopy 23(2):59–70
42. Ruiz Chica J, Medina M, Sanchez Jimenez F, Ramirez F (2004) On the interpretation of Raman spectra of 1-aminooxy spermine-DNA complexes. Nucl Acids Res 32(2):579–589
43. Hayashi T, Mukamel S (2008) Two-dimensional vibrational lineshapes of amide III, II, I, and A bands in a helical peptide. J Mol Liquids 141(3):149–154
44. Besley N (2004) Ab initio modeling of amide vibrational bands in solution. J Phys Chem A 108(49):10794–10800
45. Vargas-Hernández C (2022) Estudio de la interacción de nanoestructuras con ADN genómico (Doctoral thesis, Universidad Nacional de Colombia)
46. Guzman-Embus DA, Orrego Cardozo M, Vargas-Hernandez C (2013) Genomic DNA characterization of pork spleen by Raman spectroscopy. J Appl Phys 114(1):1–8
47. Londoño-Calderón CL, Vargas-Hernández CC, Jurado JF (2010) Desorption influence of water on structural, electrical properties and molecular order of vanadium pentoxide xerogel films. Rev Mex Fıs 56(5):411–415

Bibliography

1. Seo C, Cheong HY, Lee SH (2008) Color change of V_2O_5 thin films upon exposure to organic vapors. Solar Energy Mater Solar Cells 92:190–193
2. Mancic1 L, Marinkovic Z, Vulic P, Moral C, Milosevic O (2003) Sensors 3:415–423
3. Barth TFW, Posnjak EZ, Krist F (1932) 82(325):8
4. Musto P, Abbate M, Ragosta G, Scarinzi G (2007) A study by Raman, near-infrared and dynamic-mechanical spectroscopies on the curing behaviour, molecular structure and viscoelastic properties of epoxy/anhydride networks. Polymer 48:3703–3716
5. Nyquist R (2001) Interpreting infrared, Raman, and nuclear magnetic resonance spectra. Academic Press
6. Overbeke E, Devaux JL, Carter RJT, McGrail PT, Carlier V (2001) Raman spectroscopy and DSC determination of conversion in DDS-Cured Epoxy Resin: application to Epoxy-Copolyethersulfone blends. Appl Spectrosc 55:540–551
7. Riegel B, Kiefer W, Hofacker S, Schottner G (2000) FT-Raman spectroscopic investigations on the organic crosslinking in hybrid polymers. Part I: model reactions of simple epoxides. Appl Spectrosc 54:1384–1390
8. Aust JF, Booksh KS, Myrick ML (1996) Novel In Situ probe for monitoring polymer curing. Appl Spectrosc 50:382–387
9. Chike KE, Myrick ML, Lyon RE, Angel SM (1993) Raman and near-infrared studies of an Epoxy Resin. Appl Spectrosc 47:1631–1635
10. Don TM, Bell JP (1998) Fourier transform infrared analysis of polycarbonate/epoxy mixtures cured with an aromatic amine. J Appl Polym Sci 69:2395–2407
11. Farquharson S, Carignan J, Khitrov V, Senador A, Shaw M (2007) Development of a phase diagram to control composite manufacturing using Raman spectroscopy. Pers Commun
12. Riegel B, Kiefer W, Hofacker S, Schottner G (2002) FT-Raman spectroscopic investigations on the organic crosslinking in hybrid polymers Part II: reactions of Epoxy Silanes. J Sol-Gel Sci Technol 24:139–145
13. Londoño-Calderón CL, Vargas-Hernández CC, Jurado JF (2010) Desorption influence of water on structural, electrical properties and molecular order of vanadium pentoxide xerogel films. Rev Mex Fis 56(5):411–415
14. Kumaresan P, Moorthy Babu S, Anbarasan P, Crystal MJ (2008) Growth 310:1999–2004
15. Ferroelectrics, special issue on KH2PO4-type ferro- and antiferroelectrics 71–72 (1987)
16. Blinc R, Dimic V, Lahaynar G, Stepisnik J, Zumer S, Vene N (1968) J Chem Phys 49:4996–5000

C. Vargas Hernandez, *Introduction to Raman Spectroscopy and Its Applications*,
https://doi.org/10.1007/978-3-031-77551-2

17. Jurado JF, Vargas-Hernández C, Vargas RA (2012) Raman and structural studies on the high-temperature regime of the KH_2PO_4–$NH_4H_2PO_4$ system. Rev Mex Fis 58:411–416
18. Vargas-Hernández C, Almanza O, Jurado JF (2009) EPR, μ-Raman and crystallographic properties of spinel type $ZnCr_2O_4$. J Phys: Conf Ser 167:012037
19. Kumaresan P, Babu SM, Anbarasan PM (2008) Thermal, dielectric studies on pure and amino acid (L-glutamic acid, L-histidine, L-valine) doped KDP single crystals. Opt Mater 30(9):1361–1368
20. Orani D, Ricci A, Quagliano LG, Sobiesierski Z (1999) Physica B 262(1):775–778
21. Goerigk G, Bedel E, Claverie A, Herms M, Irmer G (2002) Mater Sci Eng B91–92:466–469
22. Leycuras A, Carles R, Freundlich A, Neu G, Verie C (1988) In: Materials research society symposium proceedings, vol 116, pp 251–254
23. Jurado JF, Vargas Hernández C, Sánchez JE, Racedo Niebles F (2010) Studies of strain in heterostructures GaAs/GaAs, GaAs/GaAs and GaAs/GaAs by spectroscopy μ-Raman. AIP Conf Proc 1267:1180–1181
24. Nakamoto K, Tsuboi M, Strahan GD (2008) Drug-DNA interactions: structures and spectra. Wiley
25. Thomas G Jr, Benevides J, Overman S, Ueda T, Ushizawa K, Saitoh M, Tsuboi M (1995) Polarized Raman spectra of oriented fibers of A-DNA and B-DNA: anisotropic and isotropic local Raman tensors of base and backbone vibrations. Biophys J 68(3):1073–1088
26. Maquelin K, Kirschner C, Choo-Smith LP, Van den Braak N, Endtz HP, Naumann D, Puppels GJ (2002) Identification of medically relevant microorganisms by vibrational spectroscopy. J Microbiol Methods 51(3):255–271
27. Muntean C, Halmagyi A, Puia M, Pavel I (2009) FT-Raman signatures of genomic DNA from plant tissues. Spectroscopy 23(2):59–70
28. Ruiz Chica J, Medina M, Sanchez Jimenez F, Ramirez F (2004) On the interpretation of Raman spectra of 1-aminooxy spermine-DNA complexes. Nucl Acids Res 32(2):579–589
29. Hayashi T, Mukamel S (2008) Two-dimensional vibrational lineshapes of amide III, II, I, and A bands in a helical peptide. J Mol Liquids 141(3):149–154
30. Besley N (2004) Ab initio modeling of amide vibrational bands in solution. J Phys Chem A 108(49):10794–10800
31. Vargas-Hernández C (2022) Estudio de la interacción de nanoestructuras con ADN genómico (Doctoral thesis, Universidad Nacional de Colombia)
32. Guzman-Embus DA, Orrego Cardozo M, Vargas-Hernandez C (2013) Genomic DNA characterization of pork spleen by Raman spectroscopy. J Appl Phys 114(1):1–8

Index

A
Anti-Stokes Raman, 122

B
Bone, 228

C
Calcite, 222
Cellulose, 223
Classical theory of the Raman effect, 121
Concept of phonons, 117

D
Detectors, 199
Diamine, 224

E
Elastic scattering, 119
Electric polarization, 12
Electromagnetic spectrum, 8
Electromagnetic wave, 3, 8, 9
Energy of the photon, 4
Enhancement factor, 205–208
Epoxy resin, 224
Extinction coefficient, 10

F
Fourier Transform Infrared (FTIR), 195, 196, 198
Free electron gas model, 144

G
Gallium Arsenide, 219, 221
Generalized matrix, 110
Group theory, 161

H
Hair, 225
Harmonic crystals, 81
Historical review of the Raman spectroscopy, 121

I
Inelastic scattering, 119
Inorganic materials of Raman spectroscopy, 212

K
KDP, 218

L
Lagrangian equation, 81
Localized surface plasmons, 138, 150

M
Matrix method, 101
Maxwell's equations, 141
Metallic nanostructures, 204
Molecular symmetry, 161
Monochromator, 200

C. Vargas Hernandez, *Introduction to Raman Spectroscopy and Its Applications*,
https://doi.org/10.1007/978-3-031-77551-2

N
Normal coordinates, 82

O
Optical parameters, 3

P
Penetration length, 14
Perturbation theory, 53, 56
Photon, 4
Planar equilateral triangle, 87
Plasma frequency, 146
Plasmon, 137, 139, 141, 158
Polarizability, 122, 127, 129, 132, 133
Polarization, 5, 145
Polyethylene Oxide (PEO), 232
Principle of SERS, 203
Principle of the Raman technique, 197

Q
Quantum theory of the Raman effect, 130

R
Raman-active, 126
Raman cross section, 129
Raman scattering intensity, 205
Raman spectra of some materials, 211
Raman spectroscopy, 119, 121, 194, 195
Raman technique, 193
Rayleigh scattering, 119, 121
Rearrangement Theorem, 167
Reflectance, 11
Refractive index, 9

S
Selection rules, 125, 161
Sellmeier, 12
SERS intensity, 206
SERS spectroscopy technique, 203
SERS theory, 137
Silicon, 212
Snail Shell, 230
Stark effect, 59
Stokes Raman, 122
Stokes scattering, 123, 125
Styrofoam, 233
Symmetries, 188
Symmetries and quantum mechanics, 185
Symmetry elements, 168

T
Theory in metals, 141
Time-dependent perturbation theory, 63
Time-independent perturbation theory, 53

V
Vanadium oxide, 216

W
Water, 231
Wave-particle duality, 5

Z
Zinc Chromite, 217
Zinc Oxide, 213, 216